Y0-AAA-352

MATH.
STAT.
LIBRARY

The Numbers Game

MATH.
STAT.
LIBRARY

The Numbers Game

USES AND ABUSES OF MANAGERIAL STATISTICS

Robert S. Reichard

Senior Economics Editor, Purchasing Week
Member of Business Research Advisory Committee,
Bureau of Labor Statistics
Member of the Faculty, New York Institute of Technology

McGRAW-HILL BOOK COMPANY

New York   St. Louis   San Francisco   Düsseldorf
Johannesburg   Kuala Lumpur   London   Mexico
Montreal   New Delhi   Panama   Rio de Janeiro
Singapore   Sydney   Toronto

MATH-STAT.
LIBRARY

MATH-STAT.

THE NUMBERS GAME

Copyright © 1972 by McGraw-Hill, Inc. All Rights Reserved. Printed in the United States of America. No part of this publication may be reproduced, stored in a retrieval system, or transmitted, in any form or by any means, electronic, mechanical, photocopying, recording, or otherwise, without the prior written permission of the publisher. *Library of Congress Catalog Card Number* 77-172032

07-051776-2

1234567890 MAMM 754321

This book was set in Intertype Baskerville by The Maple
Press Company, and printed and bound by The Maple
Press Company. The editors were W. Hodson Mogan
and Linda B. Hander. The designer was Naomi Auerbach.
Teresa F. Leaden supervised production.

*To Haze, Ali, and Peter—with an added note of thanks
to Dorothy "Bonnie" Lewis*

MATH.-
STAT.
LIBRARY

HA29
R371

MATH.
STAT.
LIBRARY

Contents

1759

Preface

To say that times have changed is to state the obvious. Yet all too often management, beset with daily operating pressures, has failed to keep pace with the ever-increasing flow of statistical data designed to put business on a more solid, scientific footing.

In a sense, this is the age of numbers. An increasingly complex society—with its automated equipment, its computers, and its need for fast, accurate, and pertinent information—dictates more and more emphasis on facts, figures, and mathematical documentation.

Indeed, the entire foundation of modern business rests on the quantification of information (statistics). Clearly the old seat-of-the-pants, intuitive technique has become next to useless in this kind of economic climate. The problems are too complex, the stakes are too high, and the competition is too savvy. Survival today virtually dictates a systematic marshaling of facts and figures—for only this kind of approach can lay the ground-work for sound decisionmaking, planning, and control.

All this, in turn, has raised a number of new management headaches. For one, how can today's busy executive possibly hope to keep *au courant* with the ever-increasing number of complex new techniques and approaches coming off the statistical drawing boards? Second, where is he going to find enough time even to thumb through the heavy flow of quantitative information that crosses his desk each day in a seemingly unending stream? Last but not least, there's the crucial problem of evaluation: the need to separate fact from fiction—the useful from the useless and misleading.

Unfortunately, there are no simple answers. But this much seems clear: The busy and harried manager can hardly be expected to know as much as the highly trained statistical analysts who generate the information. On the other hand, he must know enough to make quick dollars-and-cents decisions on the basis of such data. It is to this latter task that *The Numbers Game: Uses and Abuses of Managerial Statistics* is addressed.

Put another way—the objective here is to examine how and under what conditions specific statistical tools and techniques may or may not be applied. Equally important, emphasis is placed on the dangers of misinterpretation—for the statistical graveyard is replete with examples of how distortions and misunderstandings have led to erroneous decisions.

The Numbers Game, then, is intended essentially as a comprehensive handbook of statistical do's and don'ts—one that enumerates all the classical pitfalls and yet is at the same time wide enough in scope to permit the development of a workable system of critical quantitative evaluation.

To achieve this goal, it has been necessary to approach the subject of statistics from a somewhat different angle. The book's uniqueness lies in the fact that it stresses a combination of factors, including the following.

Functional understanding. The emphasis is on how, when, and where to use or not use a particular technique or approach. Formal mathematics and formulas are used only sparingly—and only when absolutely necessary to understanding or interpretation. This it not to say, of course, that accuracy is being sacrificed on the alter of simplicity. Indeed, great pains have been taken not to water down any important statistical concept. A simple formula or two may be presented to illustrate a principle on the verity or spuriousness of a particular approach. But as implied above, it is the functional aspects of the use and misuse of figures that get the lion's share of attention.

Business orientation. Great pains have been taken to address the book to the nontechnical corporate executive. Thus the accent is never on knowledge for knowledge's sake—but rather on down-to-earth applications involving the workaday problems of the corporate world. Moreover, the book is faithful to the language and flavor of the business community—pointing up what is important to the decision maker rather than to the statistician in his ivory tower. Finally, to inject the crucial note of reality, the stress is on practical, action-oriented problems—most of them culled from actual industry experience.

Wide-angle range. Rather than zeroing in on one or two particular aspects of managerial statistics, *The Numbers Game* takes a more catholic view, touching upon as many different areas as it is possible to cover in a book of this length and scope. Thus the subject matter ranges from such seemingly basic yardsticks as averages and percentages all the way to such complex space-age approaches as linear programming and input-output analysis. The hope is that when the reader has completed the book, he will have at least a nodding acquaintance with virtually every useful technique currently used by business decision makers.

Built-in selectivity. Each individual subject has been treated as an independent entity, able to stand on its own two feet, and there is a minimum of time-consuming cross-referencing. Thus the busy executive who wants a fast rundown on, say, correlation is no longer forced to become an expert on averages, statistical dispersion, and regression analysis. By the same token, expertise on correlation is not deemed a prerequisite for a basic understanding of mathematical models.

Dollars-and-cents practicability. Techniques are evaluated on more than just statistical efficacy. Payoff is deemed equally important. Thus, each dollar spent on upgrading business intelligence must be justified in terms of the dollars and cents it will eventually pull into company coffers via improved planning and operating efficiencies. Accuracy without payoff is a luxury that few firms can afford these days.

An added plus stems from the fact that the book has been carefully designed to provide an overview of current management practices and policies. This, in turn, should provide some useful yardsticks for gauging company performance—and, one hopes, should spark new ideas for getting still better mileage out of industry's statistical investment.

Finally, *The Numbers Game*—because it assumes little in the way of mathematical sophistication—should also prove useful in both executive training programs and introductory college courses in managerial statistics. Indeed, much of the material has been used by the author in a beginning course in business statistics—with gratifying results. Students were more interested, earned higher marks, and—most important of all—walked out of the course with a real knowledge and understanding of what statistics is all about.

But whether used in industry or school, the book hopes to hammer home one vital point: It is no longer possible to take the serenely optimistic view that the future will automatically take care of itself. It's just too risky—a long shot that rarely pays off. If nothing else, the illustrations contained in *The Numbers Game* make it amply clear that only through better statistical intelligence is it possible to turn a threatened loss into a profit—and a marginal enterprise into a going and growing concern.

Robert S. Reichard

ONE

Statistics in Today's Business World

The dictionary defines statistics as a science involving the collection and classification of facts on the basis of relative number or occurrence. But in the real, workaday world this simple and seemingly straightforward definition can mean many different things to many different people.

To some, statistics means little more than the row upon row of dry figures on sales, income, profits, wage levels—or whatever else is being tabulated at the time.

To the professional statistician, the term implies much more—including the use of computer-oriented mathematical formulas which enable one to better understand and predict the complex world in which we live.

To still others—mostly laymen—statistics is nothing more than a handy tool for lying with numbers.

In a sense, all three interpretations have some validity, for statistics (1) do require the seemingly endless collection and tabulation of numbers, (2) do lend themselves to algebraic manipulation and projection, and (3) can be used to either illuminate or deceive.

Even the most avid proponent of the statistical approach would agree that the chances of distortion, manipulation, and outright deception have increased substantially as a result of the numbers explosion of the past few decades. (Who hasn't, for example, at one time or another been fooled or misled by statistical legerdemain?)

On the other hand, it is equally true that today's fantastic progress—in business as well as in the physical and social sciences—is due in large part to the collection, tabulation, processing, and analysis of quantitative data.

Thus, when toting up the balance sheet, the problems are indeed a small price to pay for such impressive statistical dividends. For thanks to quantitative analysis of business data, exploration of the future is no longer equated with fortune-telling or with the contemplation of crystal balls. Instead there is a growing awareness that a great deal can be said about future trends in terms of probability, and moreover that through proper planning we can exert considerable influence over these probabilities.

In a sense, the future is no longer viewed as unique, unforeseeable, and inevitable; there are instead—it is realized—a multitude of possible futures, with associated probabilities that can be estimated and, to some extent, manipulated. But despite all the clear-cut advantages to the use of quantitative analysis of business data, the antistatistical school remains surprisingly vocal—numbering among its advocates a significant number of businessmen and other professionals who should know better. Remarks such as "You can prove almost anything you want with numbers" or "Figures don't lie; liars figure" didn't spring up out of thin air.

Too many businessmen, politicians, scientists—indeed anyone who deals with the quantification of data—have used statistics to prove their point rather than to provide an unbiased picture of the real world. Who hasn't at one time or another been tempted—to be sure, in the heat of argument—to take some data out of context in order to prove a crucial point?

Then, too, who hasn't met the type of salesman who brags about having doubled his sales in one year—conveniently forgetting the fact that the introduction of a new product by his firm has allowed virtually all the firm's salesmen to rack up similar large gains?

But to summarily dismiss the statistical approach because of this and similar distortions is simply poor logic. It's like arguing for an end to efficient automated production because it can lead to problems such as technological unemployment, strikes, etc. It boils down to little more than a "Let's throw the baby out with the bath water" attitude.

In any case, dishonesty isn't the only criticism leveled at the statistical discipline. Many like to point up the lack of significant meaning in some quantitative measures. Usually those who demean the statistical approach in this fashion like to zero in on the concept of averaging—hence such remarks as "The average American shaves half his face and powders the other half" or "A man on the average is comfortable when he has one foot in boiling water and the other in a pail of ice water."

These barbs, some of them born out of unhappy experience, play up the ludicrous. They ignore the great deal of useful information that such simple yardsticks as the average can provide. They also give short shrift to the fact that averages are the key building blocks for many of our sophisticated statistical approaches.

At still other times, charges of dishonesty are basically unwarranted—simply because distortion is sometimes only in the eyes of the beholder. Thus, what seems to be misrepresentation to one observer is seen as pure truth by another. This kind of misunderstanding sometimes revolves around the "Is the bottle half full or half empty?" dilemma.

A simple example best illustrates the problem. In 1968 steel buyers built up a substantial 13-million-ton inventory hedge in anticipation of a strike which never materialized. Once a new pact was negotiated, the problem was to dispose of the excess metal as quickly as possible. After three months, buyers succeeded in just about halving their previous hedge.

Those who were satisfied with this stock-paring pace accentuated the positive—namely, that more than 6 million tons had been whittled off inventory totals in an extremely short period of time.

Other analysts, however, preferred to point out that despite paring progress, some 6 or 7 million tons of excess metal was still hanging over the market.

Both points of view were correct, both equally valid. Which one to accept depended on one's basic interest. From the buyers' point of view, knocking some 6 million tons off their shelves in only three months was no mean feat—and was certainly worth crowing about.

On the other hand, steel producers, who had to sell steel and plan production schedules, could hardly afford to neglect the second view, for it emphasized the fact that it might be another few months before they could count on normal sales patterns.

In short, depending on particular needs, the steel bin was either half full or half empty.

Semantics, too, can lead to statistical disagreement. What is defined one way by one analyst can be viewed in a completely different light by another. The squabble over desegregation is a case in point. Thus in 1970 the federal government stated that 94 percent of the Deep South public school system had been officially desegregated. But upon further examination it was found that official integration was often accompanied by de facto segregation.

In short, one's estimate of progress in desegregation depended on one's definition of the word "desegregation." Among the many indirect methods used by the South at that time to circumvent official desegregation were (1) the administering of ability and achievement tests to determine class makeups, a technique which resulted in virtually all black or all white classes in some schools; (2) the trek to private schools, which left desegregated schools predominantly black; (3) the gerrymandering of school districts to keep the races apart; and (4) allowing students to choose their classes (they nearly always opted to stay with their own kind).

To be sure, many Southern school districts actually were desegregating at the time. Indeed, even the most pessimistic of civil rights leaders admitted to some progress. What they objected to, though, was the fact that the listing of the 94 percent desegregation figure was misleading—that it was convincing many well-meaning people that the integration job in the Deep South was nearing completion.

The history of semantic differences in statistics is a long one. Such seemingly clear-cut yardsticks as cost, depreciation, or profit can often mean different things to different people. That's why it's so important to define one's terms.

But semantic differences hardly explain all problems. Sometimes distortion of data exists but is unintentional—due, in part, to the user's lack of sophistication rather than to any premeditated intention to misinform. Take the current popularity of government releases such as the monthly report on changes in the cost-of-living index.

Because of widespread reporting in the press, the government price measure has become a household word. The trouble is, however, that these reports are analyzed in many cases by people who don't have the slightest statistical qualification to interpret the results.

Indeed, the way the price trend is reported in the press could lead to the conclusion that it is fantastically erratic. Let the cost of living rise 0.1 percent, and some are led to believe that runaway inflation is upon us. Let a few prices decline the next day, and these same interpreters would have us believe that the nationwide inflation problem has finally been licked.

Aside from the lack of sophistication, this erratic treatment is often forced upon the press—primarily because each story, to be justified, must be newsworthy. Under this criterion a change of, say, 0.1 percent in Uncle Sam's price index would not be worth reporting without a press buildup—because it signifies little change from the previous month.

Nor does the government always come out lily-white. As Arthur M. Ross, a former Commissioner of Labor Statistics, recently put it:

> Statistics consists of technical procedures quite independent of content or purpose. I found that most government statisticians are principally concerned with techniques, which have greatly improved in recent decades. But their outlook is often too narrow to encompass the larger role of numbers in public life.
>
> Like horses who obediently pull a wagon over a cliff, they exercise great skill in producing numbers but have little sophistication concerning their use and misuse. Although statisticians like to think that they are constantly warning policy-makers against misuse, what they usually warn them about is the limited sample or the possible range of error, rather than the one-dimensional quality of the statistics themselves.

In short, Ross's concern is not with statistics but with the abuse of statistics. His conclusion: "It is human ignorance, indolence, and incuriosity which permit statistical data to be perceived as objective verities rather than as the shadowy hints and clues they most often are."

In another vein, day-to-day reports can also be confusing because of the vast complexity of the typical American business enterprise. Not everything moves up and down in tandem. More often than not, one key indicator of, say, a firm's sales potential can be moving up while another is heading down.

Take the case of an appliance maker trying to get a handle on potential sales. Normally he would gauge the outlook on the basis of consumer incomes and housing starts. But it is often true that consumer incomes continue to rise when housing starts are falling. The problem: How to interpret?

The answer is not an easy one. But obviously it requires detailed study of the interaction of these two sales determinants—and then measuring their combined effect on future appliance purchases.

The technique used is *multiple correlation*—and will be discussed in more detail in Chapter 7. But the important thing to keep in mind—in this as well as in any other case involving statistical interpretation and analysis—is the necessity of developing action-oriented guidelines—yardsticks which can evaluate and make sense out of seemingly conflicting intelligence.

Such guidelines are also useful in signaling some basic shifts in the business climate. A recent case involving retail sales and inventories comes to mind. Sales, after a period of lackluster performance, had started to move up smartly. After a few months of this, businessmen started laying in heavier inventories on the assumption that the good performance would continue. But this simplistic extrapolation approach to market evaluation tended to ignore such key sales guidelines as (1) an already accomplished bounce back to normal sales/income ratios and (2) a deteriorating consumer liquidity position.

The unwitting decision to ignore these factors—and to accept instead the rosy day-to-day sales reports—led to a serious miscalculation and an unwanted inventory buildup. More important, a lot of expense and heartache could easily have been avoided by the proper use of available market clues.

An equally pressing statistical problem stems from the fact that there are many who accept any quantification of data as gospel. This even occurs among professional college graduates who know about the difficulties involved in such common data collection processes as sampling. And yet, knowing this, these worshipers of statistics stand in rapturous awe of each new bit of quantified information that comes across their desks.

This is somewhat akin to our tendency to accept anything we see in print as the gospel truth—simply because it has been printed. Let the government issue a preliminary figure on retail sales or machinery orders, and some people are ready to accept it as revealed wisdom. Indeed, even a suggestion of criticism verges upon irreverence.

Finally, when the power of massed rows and columns of numbers is buttressed by the force of the computer, criticism passes the irreverence stage and borders on rank heresy.

This problem is by no means rare—especially in the executive suite.

Having shelled out several million dollars for a new computer, many businessmen are loath to admit that it can do any wrong.

Part of this, of course, is due to lack of sophistication about what the computer can and cannot do.

Computers can't do the thinking for you, nor can they create something out of nothing. More specifically, it should be immediately apparent (but many times isn't) that the GIGO—or garbage-in, garbage-out—approach to quantitative analysis is equally applicable to counting on one's fingers or using an expensive new computer.

Without good input, good output is impossible, no matter how some businessmen might try to rationalize otherwise.

That's not to say, of course, that every input figure must be 100 percent accurate. This would be neither feasible nor practical—for perfection in statistics as in almost any other type of endeavor is an impossible goal. Thus, while the United States has the best market reporting system in the world (as measured by quantity, quality, and timeliness), users are always pointing to many important gaps and to ways in which reporting could be improved.

Even so, the potential for statistical evaluation and analysis is great. There are enough good data generated—by government, by trade associations, and by individual firms—to permit a substantial improvement in almost any company's operations.

In short, the goal of statistical analysis is to get the maximum mileage out of what is available. With literally millions of dollars of a firm's money often riding on a single decision, the effort to make use of somewhat less-than-perfect data becomes almost mandatory.

Take the recent dilemma of a primary metal producer. He knew that his sales depended in part on the prices of competing imports. But no such price measure was available.

The firm solved the problem by compiling a list of price quotations from leading exporting nations. Then it weighted these reports by the amount of metal coming in from each of these countries. To be sure, the resulting yardstick was only a makeshift average—and by no means perfect. But it proved reasonably accurate in explaining shifts in company sales over a period of time. More important, it provided an important guide to future production and sales policies.

The above example illustrates still another point: Without the proper quantitative market data, attempting to make decisions on such key questions as sales potential, capital spending, or inventory policy becomes extremely risky.

It used to be that a businessman could afford to make mistakes and could get away with them—primarily because his counterparts in competing firms could be counted on to make equally costly errors. But this is not true today.

The majority of companies today are staffed with a corps of experts whose sole function is to analyze quantitative data. With the development of computers and new sophisticated techniques, these technicians are not likely to go off the beam very often or very far.

With all this statistical effort, the typical company today has been able to reduce substantially the probability of error. Decision making no longer involves the pure gamble of, say, poker playing or horse-race betting. True, there is still a chance of erring, but the odds are becoming more and more stacked in the businessman's favor.

To sum up, the collection and analysis of quantitative data are basic and inescapable responsibilities of business management. The systematic marshaling of facts and judgment for gauging future company prospects is essential for sound decision making, planning, and control.

How then can a successful statistical program be built up—one that embraces the middle- and upper-management echelons as well as the soldiers on the statistical firing line?

The basic problem, of course, is to build up the level of competence and sophistication within the organization. And it is to this basic goal that this book is addressed. But before going into the specific do's and don'ts of management statistics (the subject of the following chapters), it is important to recognize some of the basic organizational and procedural problems that underlie any utilization of quantitative data for the upgrading of corporate performance.

The following, then, are some of the broader-gauge questions that any quantitatively oriented management will want to consider. They cut across the small and the large, makers of durable and nondurable products, and producers of consumer and capital goods. Indeed, any outfit set on upgrading the quantity and quality of its marketing intelligence is almost sure to run into the following common problems.

The Problem of Newness

For many, rigorous quantitative analysis is a new approach. For others the newness applies to a particular technique rather than the overall approach. In either case, however, there is usually an absence of a firm body of statistical dogma—at least as it applies to any given firm. This, in turn, makes quantitative analysis a wide-open field—not subject to the safe and well-worn paths of other management areas such as labor relations or corporate finance. This creates problems—but at the same time offers major opportunities for intelligent innovation.

Relative newness of statistical emphasis also creates some personnel headaches. Many in important decision-making positions have received only a cursory kind of introduction to statistics in college—and that probably occurred at least 10 years ago.

These people, as bright and adaptive as they are, cannot be expected to jump into this new and fairly complex area in a matter of weeks or even months. As is true of any new tool, time and effort are required to ensure the proper use of quantitative analysis.

In short, the building up of a smoothly functioning statistical machine just doesn't happen overnight. It requires a considerable amount of planning and often juggling of personnel.

It has been said that a good regression analyst—one trained in the art of gauging the effect of one business variable upon another—needs upward of three years of practical experience, in addition to the traditional college training. The chances of finding such a man already on the staff are indeed remote.

Another aspect of the newness problem deserves some comment. Specifically, the use of numbers to solve business problems often means treading on untested ground. It is by no means certain, for example, that what has worked in the classroom will work in the real world.

Thus, in an academic setting, the data are always "given." The basic problem is then one of processing and interpreting.

Not so in the real world when a new program is being initiated. More often than not, the basic data have to be researched. And in some cases they even have to be custom-tailored to meet the stringent requirements of a new mathematical approach.

Then, too, the real world contains a great many more variables than any hypothetical classroom problem. In almost every case a relatively complex adaptation of the theoretical approach will be needed to obtain any meaningful results.

Theory, for example, may tell us that income and price determine sales volume. But it doesn't answer the question of "what income." Is it gross dollar income, take-home pay, or real income (income with the inflation factor filtered out)? The answer may well be different for different products.

Similar problems involve the influence of price. Is it the price of the commodity in question, the price of the nearest substitute, the price of the nearest three substitutes—or perhaps a combination of all these factors? Only the marketplace can give the precise answer.

Newness also brings up organizational problems. If a new research function is being set up, under what department should it be placed? What are the chains of command? How is it to be integrated with other staff functions of the organization?

Internal politics often compound these setup problems. If marketing and sales are vying for power, it is going to be extremely difficult to convince either one that the critical sales forecasting function should be put in the other one's bailiwick. These are over and above the normal organizational and cordination problems discussed on page 20.

The Rapid Expansion of Statistical Analysis

Today, standard operating procedures in the handling and analysis of data vary significantly from those used yesterday. And those of tomorrow will in all probability vary just as significantly from today's. The point is that technical developments make it virtually impossible to stand still—even where a program is working out quite satisfactorily. The problem here is many-faceted.

For one thing, there are the competitive aspects. In today's dog-eat-dog world no firm can afford to rest on its laurels. If one firm comes up with a new scientific approach, say, for pointing up equipment needs, other firms have to develop similarly effective techniques of their own. Underestimating or overestimating future capital requirements in terms of market or modernization pressures can mean a sharp competitive setback for a firm.

A few years ago, for example, a large electrical equipment manufacturing company unwisely put too many of its capital spending eggs in the nuclear energy basket. Meantime, a sharp rise in consumer appliance demand occurred. But the company, because its capital was earmarked for other purposes, was unable to take advantage of this—and was forced to accept a temporary drop in its appliance market share.

Sometimes a whole new discipline must be added to the arsenal of tools used to plan the future. Technological forecasting is a case in point. Today, a rapidly changing state of the arts virtually dictates a systematic and formalized approach to predicting the nature and timing of new innovations and developments.

Admittedly a difficult task, technological forecasting has fantastic potential. Forecasts, even if of only limited accuracy, can (1) aid in planning long-run R&D programs, (2) identify opportunities for diversification and expansion, (3) establish more accurate product time-tables, (4) pinpoint potential customers' needs and markets, and (5) spotlight competitive threats and potentially obsolete product lines.

In any case, this new discipline, based on quantitative techniques rather than opinion, can be a powerful tool for improving corporate long-range planning.

Sometimes the pressure to expand statistical horizons comes from internally rather than externally generated sources. Let any mathematically oriented approach prove successful, and it's a pretty good bet that the personnel responsible will start opting for even more ambitious approaches. And so they should.

But whatever the motivating force, the problems of statistical expansion are almost as great as those involved in setting up an original program. This is certainly true in the generation of basic data. For what was adequate for a less ambitious program suddenly becomes woefully inadequate.

Take the example of the machinery firm that had been basing its forecast of future activity on the cash flow of all United States manufacturing firms. The approach had worked out fairly well, but then the firm decided to analyze its sales on both product and geographic lines. This meant that the aggregate cash flow figures weren't enough. The cash flow data had to be similarly cross-classified if any meaningful results were to ensue. Unfortunately, this wasn't an easy task—it required considerable money, time, and effort. Nevertheless, the ultimate payoff—in terms of more accurate projections—eventually justified the move.

Expanding marketing programs also usually implies the utilization of some of the more sophisticated techniques that have been developed only recently. This, in turn, creates a staying-in-touch problem.

Corporate staffs thus have the added task of remaining *au courant* as regards the latest approaches. This is not always easy. New computer techniques to test new product potential aren't something one is likely to find in the average college textbook—because of both newness and the inherent complexity involved. Nevertheless, it has become a virtual must for large corporations to keep a weather eye out for new developments in this area. And with good reason: New products often account for upward of 10 percent of a firm's sales in any given year.

The Need to Sell the Function

Like any staff function (as opposed to line function), quantitative market analysis is likely to encounter inertia or sometimes even outright hostility on the part of some old-line managers. Their basic argument: "We fared pretty well without this new mathematical gobbledygook for over 20 years, so why should we spend time and money now for a change?"

There's no pat answer to this one—except to stress that business today is not the same as it was 20 years or even 10 years ago. In our rapidly changing technological society, there's too much at stake to trust to the old seat-of-the-pants approach.

Fortunately, this problem is now limited mostly to small and medium-sized firms. Virtually every corporate giant has jumped onto the planning bandwagon—with billions earmarked each year for statistical intelligence of one type or another.

The usual argument of smaller firms—namely, that quantitative market analysis is too expensive—just doesn't hold water. Indeed, with change so rapid, the question should be not whether you can afford to use a statistical approach, but whether you can afford not to use it.

Moreover, not every program is prohibitively expensive. Even in the computer area, renting time on someone else's equipment or purchasing one of the inexpensive minicomputers now hitting the market offer practical ways of embarking on sophisticated statistical programs at relatively low cost.

The money-saving potential of these more sophisticated techniques can be demonstrated in any number of ways. For example, if a firm that usually carries $100 million in inventory can reduce it by only 5 percent (thanks to new economic order quantity techniques), the gains can be surprisingly high.

If, for example, you accept the fact that carrying inventories currently costs about 2 percent a month, or 24 percent a year, then it's easy to calculate the potential savings. Specifically, the 5 percent reduction in inventories over the year will result in a savings of nearly $1¼ million dollars to the firm (24 percent of the $5 million inventory decline). Added advantages, of course, accrue from the fact that the company is freeing previously tied-up money for more remunerative purposes.

The Unsettling Factor of Change

This is one family of headaches that no firm can afford to ignore. Any statistical report is based on certain logical assumptions. Unfortunately, economic, political, technological, and market factors are prone to fast change. This in turn calls for constant vigilance and frequent revision when and if the ground rules change.

Chapter 14 will treat the problem of changing assumptions in more detail. But at this point certain broad-gauge caveats seem in order.

On the technological front, its important to keep in mind that a new invention can completely change the outlook and relationships. Thus, before the invention of the auto, it was fairly easy to forecast demand for the old horse and buggy. However, once the auto age was ushered in, all bets were off. Coming closer to the present, the development of plastics has completely altered the sales outlook for metals and other raw materials which compete with the man-made product.

Institutional or underlying changes in the basic way in which the economy functions must also be factored into any quantitative blueprint. One such change that comes to mind is the recent trend toward spending more for services and less for physical merchandise. Such long-run trends tend to pull a forecast one way or the other if the old relationship is maintained year after year.

On the political front, changes can be equally disrupting. The end of a war, the raising or lowering of income or excise taxes, or even a change in administration can make for some major shifts in a firm's basic production, sales, and capital spending outlook.

Market developments also inject a note of uncertainty into company plans. Let an industry leader stress safety or repairability of its products, and it's incumbent upon other producers in the industry to react similarly—causing major changes in every firm's master plan for future operations.

Sometimes a change in marketing trends can be extremely difficult to monitor. Fashion shifts are a case in point. Take, for example, the horrendous flop of the long (midi) skirt in 1970. Manufacturers, following the suggestion of fashion designers, put all their money on the acceptance of the midi. But their projections that year proved to be way off, as the consumer stuck with the old miniskirt. Result: Widespread manufacturer losses, with many outfits forced into bankruptcy.

Just how bad the situation was in 1970 can be gleaned from a Gallup poll taken at that time. An unbelievable 0 percent of women between the ages of twenty-one and twenty-nine opted for the long skirt—and only 3 percent of those between thirty and thirty-nine thought about switching to the new style. More significant, only 1 percent of those polled registered no opinion—a sign that advertising, promotion, and social pressures could influence very few fence-sitters.

Why the midi length failed in 1970 came out in subsequent interviews with women. Generally speaking, the consensus at that time was that the long skirt made fat girls look fatter, tall girls look dumpy, and older women look older. One woman even likened the midi to the Edsel—a Ford automobile that flopped miserably during the early 1960s.

What made matters worse was that the indecision engendered by the introduction of the midi that year dragged down all sales—of short as well as long hemlines. As one retailer put it: "If you have a fashion trend introduced with a lot of fanfare and it doesn't catch on, then it means you not only don't sell the new styles but also don't sell something else you could have sold. Women in such a situation are confused and not sure what is right, so they don't buy anything."

And his evaluation of the situation was later borne out. One apparel man toward year-end estimated that dress and skirt sales in 1970 fell as much as one-third from the previous year.

Another complication during this fashion debacle was the fact that the new style was introduced during a period of recession and stock market weakness. This meant that the people who had to be convinced about the midi length—those with money—were then in economic hot water.

Before leaving the subject of change—and its effect on forecasting—one other problem might be mentioned: the dynamic relationship or interaction between the various series a firm uses as basic quantitative input. Take an example from the field of sales forecasting. The price tag a product carries affects its sales, but the volume of sales, in turn, affects the price needed to make a fair return. The point to keep in mind is that there is often a close connection between the input and the output of a statistical relationship and that to ignore this dynamic relationship can sometimes lead to serious miscalculations.

Knowing Your Industry

The statistical approach by itself is meaningless—unless it can be directly related to the bread-and-butter applications of the workaday business world. Put another way: To manipulate data, one must first have knowledge of the data earmarked for manipulation.

How, for example, can one plan for production schedules of, say, autos without firsthand information about the myriad of specific factors that influence auto production? And in this case there are many—one analyst figures there are over 25 that can have a significant effect on the level of automotive activity.

How does the statistical technician find out which ones? Certainly not by reading textbooks. Yet this intimate knowledge of the industry is a must ingredient, along with the formalized statistical sophistication. In the words of an old popular song, "you can't have one without the other."

Sometimes even business-oriented statisticians—wrapped up in their own little world of numbers—can go off the beam. A recent instance in the electronics industry comes to mind. Mathematicians began making forecasts of growth lines, going 10 to 15 years out into the future. On the basis of past trend relationships, the projections seemed to be perfectly sound and consistent. But one look by the sales manager sobered up the researchers. He pointed out that their aggregate sales prediction a decade from now exceeded the probable level of the gross national product (GNP)—the sum total of all business activity in the United States.

Changing Usefulness of Yardsticks

Many statistical series which were accepted without question a decade or so ago are being increasingly questioned as to validity. In almost all cases they involve the equating of statistical fact with absolute truth. This can be best explained through a concrete illustration.

It may be possible, for example, to ascertain that Mr. X did better than Mr. Y on a work aptitude test. But to assume that this means Mr. X will be a better worker than Mr. Y is to assume the infallibility of aptitude tests—something which in most cases seems to be unwarranted.

In a similar vein, it is a fact that the United States had an extremely favorable "kill" ratio relative to that of the North Vietnamese. But the simple assumption that this meant we were winning the war in Vietnam was quickly disproved.

Finally, it is worth noting that such widely accepted measures of progress as the GNP (the market value of all goods and services produced) are coming under increasing attack as national yardsticks. Many claim that GNP growth deceives us into complacency that all is well and getting better, when in actuality things may be getting dangerously worse.

Much of this revolves around the problem that as material wealth grows, it tends to pollute and generally speed up the deterioration of the environment. Arthur Burns, Chairman of the Federal Reserve Board in 1970 summed it all up when he said he would like to see the GNP adjusted to "take account of depreciation in our environment." When there is a "proper recording of the minuses as well as the pluses," Burns predicted, "we will discover that the GNP, which has been deceiving us all along, is a good deal lower than we think it is."

As its name implies, of course, the GNP is "gross" because it disregards even the conventional kind of depreciation—the wearing out of plants and equipment for which businesses are allowed tax deductions. To be sure, the Commerce Department does publish a net national product total that is smaller than the gross, or GNP, figure, but scarcely anyone notices.

Moreover, even the net national figure does nothing to account for environmental depreciation, and according to some estimates, this amounts to as much as $30 billion to $40 billion per year. In short, the GNP tells us only of quantity and nothing of quality, neglecting to make any subtraction for the losses in amenity which are the consequences of overcrowding, pollution, transportation failures, housing shortages, and the sheer pressure people exert upon one another.

The GNP's lumping together of unrelated quantities has also come under increasing attack. Certainly, one can't tell whether the GNP's huge consumer services sector is growing because more people are gratefully paying for better medical care or because more householders are irately paying for repeated repairs on shoddy appliances. Similarly, others claim we are misguided by adding military equipment into GNP totals.

These criticisms aren't without policy implications. If pollution eroded rather than bolstered the GNP, for example, some lawmakers figure we'd long since have had such safeguards as extra taxes to cover the cost of cleaning up beer cans or oil spills. Conceivably, the nation might have been warier about foreign wars if their costs weren't routinely treated as adding to economic growth.

To sum up, after a decade spent in the spectacularly successful pursuit of a rising GNP, the United States has suddenly begun to have second thoughts about growth as a goal for national policy. The exceptional growth rates of the sixties are just about taken for granted as a model for the seventies. Yet more and more people are beginning to see economic growth itself as the basic cause of the environmental deterioration that has become one of the nation's most pressing problems.

In his book *Technology and Growth: The Price We Pay,* British economist E. J. Mishan charges that "as the carpet of increased choice is being unrolled before us by the foot, it is simultaneously being rolled up behind us by the yard." He thinks that modern societies like Britain and the United States have reached a point of diminishing returns, where the disamenities generated by increased production outweigh by any reasonable measure the value of "more cars and transistors, prepared foodstuffs, and plastic objets d'art."

Others aren't quite so pessimistic. Rather, they would change the statistical concept to take this into account. They say, in essence, that expenditures on pollution actually produce satisfactions that go with improvement in the general environment—and hence should be counted as part of the GNP. This could be done by treating environmental control outlays as quality improvements in products. Indeed, that is what the U.S. Bureau of Labor Statistics is doing with auto exhaust systems, and there is no theoretical obstacle to treating similar expenditures the same way.

Asking the Correct Questions

Because no two firms are alike, no one approach or emphasis is best for all firms. But before embarking on any major statistical program, the posing of some basic questions can at least point the way toward the proper avenues of attack.

Points to consider would include:

1. *How much money?* Obviously, a small firm can't afford a large in-plant computer-based operation. A good rule of thumb is to evaluate the amount that similarly situated companies are spending on similar programs and use this as a first approximation.

Remember, however, that the mere fact that five other firms may be spending relatively little on a particular area for statistical intelligence is no a priori reason to follow suit. All the competitive spending yardstick can do is to provide a jumping-off point or ball-park figure from which to start your own detailed calculations.

The probable saving involved is another determinant of how much money to spend on any new data gathering and processing system. Thus if electronic data processing can cut inventory and clerical costs by $1 million a year, then the outlay of several hundred thousand dollars is clearly justified—provided, of course, it doesn't result in too big a drain on operating capital.

2. How much accuracy?   Closely allied with the spending question is the companion one of accuracy. Clearly, once dollar limitations are set, a company can buy either a small statistical package with great accuracy or a larger one with not so great accuracy.

The problem is to decide what kind of trade-off is best for your own particular type of operation. To a large extent, of course, this will depend on a company's product mix. To exaggerate the point: An airplane manufacturer would necessarily have to insist on a much more highly sophisticated form of statistical quality control than, say, a maker of women's coats.

The state of the art may also play a role in determining how much precision to expect. Thus a firm introducing a new product should expect its cost and revenue estimates to be somewhat less precise than, say, those made for an established, market-tested product such as automobiles or refrigerators.

It would also be logical to expect less accuracy for individual products vis-à-vis entire product mixes. This is essentially a reflection of the "law of large numbers," which states that errors in component parts tend to cancel one another out.

3. What are the critical areas?   To a large extent the answer depends on individual company needs. But experience has shown that certain functions are particularly amenable to a quantitative problem-solving approach. These include quality control, sales forecasting, inventory management, market research, production control, capital planning, materials handling, and general distribution.

This is by no means an exhaustive list. Rather, it is an enumeration of some of those areas where big payoffs in terms of more efficient operations have been historically significant.

Sometimes quantitative analysis along product rather than functional lines may be in order. Again, company versus industry performance can provide some useful clues. If profits (the ultimate measure of performance) in some product line are well under the industry average, then clearly some sort of statistical investigation may be necessary to search out and correct the problem.

At other times communication may be the key area to focus on. More specifically, a firm may be generating ample statistical intelligence, but simply not filtering it down to these who have a need to know. Other problems involving proper organizational setup will be discussed in the next section.

4. *What outside help is available?* If an individual firm is just setting out on a new, ambitious program—or if it has a limited staff—it may be a good idea to seek the advice of appropriate trade associations or business consultants. Thus, trade associations often have their own statistical gathering services and hence are probably in a good position to advise a firm on what is practical and what is not.

Consultants can be equally useful. Indeed, some are willing to undertake entire statistical programs for individual firms. While the costs may seem high, the long-run benefits can often more than offset the basic charge. Some of the statistical work commonly performed by consultants includes (1) the working up of market surveys and (2) the generation of special marketing data not available from other sources. A consulting outfit's ready availability as a statistical sounding board is another important plus.

In still other cases, sellers of capital equipment can often provide needed guidance. Some of the big sellers of computer hardware are a pretty good source of information about how and where to use sophisticated electronic data processing techniques. These people know what the equipment can and cannot do in the area of solving complex business problems. Moreover, they are willing to spend time and effort if there is a good chance of nailing down a future sale.

In a similar vein, machine tool and other equipment makers will usually help the buyer set up a machinery evaluation program—one designed to signal when current equipment may be ready for replacement. To be sure, these people have a vested interest. But the techniques they espouse have been tested in the crucible of business experience.

5. *Does the project have multiple uses?* Other things being equal, the more areas a particular program can be applied to, the more valuable it is likely to be—and hence the more likely it is to get the go-ahead sign from top management. Thus it may be difficult to justify, say, an inventory coding system on the basis of facilitating releases to the production department. But if it can be shown that the proposed system would also facilitate computer ordering, reduce the danger of stockouts, and reduce overall inventory levels, then the project has much more going in its favor.

Similarly, a statistical program that can eventually become a building block for a more ambitious one is preferable to one that is locked into a single specific application. Thus, a few years ago computerized testing began to catch on. One of its big talking points—in addition to reducing labor costs and improving product reliability—was that the system could eventually be integrated into a far more comprehensive management information system.

6. *Is the suggested approach the appropriate one?* The question sounds naïve, but many firms embark on a particular project under the impression that it will solve their problems, when in fact it often cannot.

Consider the area of capital planning. Basing future capital needs on projected sales dollars could be misleading because gains may be due to price inflation rather than to any increase in physical volume.

In the same area, deciding on what type of equipment to buy on the basis of, say, payback of original investment can be equally misleading. For the value of any investment depends on the revenue returned over its entire life—not on how soon the original investment is returned.

To take a simpler illustration: The comparing of one month's sales with those of the previous month can lead to some pretty strange results if seasonal influences are strong. If this is the case, then year-to-year comparisons or the seasonal adjustment of the data may be called for.

Indeed, much of this book is dedicated to separating out (1) the seemingly correct approach from (2) approaches that are soundly based on statistical theory.

Organization and Coordination

No matter how good the actual techniques and approaches, a statistical program is doomed to mediocrity without proper procedures for implementation, monitoring, and evaluation. The problem is particularly acute in firms where quantitative research has increased sharply over the past few years. Unless an organizational setup is developed to accommodate the spate of new programs, inefficiency and duplication are bound to drain off a good portion of the potential benefits.

Again the trouble is that no hard-and-fast rules can be given. That's because what is right for, say, a highly centralized firm may turn out to be woefully inadequate for a diversified one. In a similar vein, a consumer-oriented operation (where style and buyer whims play a major role) could require a somewhat different setup from that needed by a firm whose activity is geared to more predictable industrial markets.

But no matter what the type of firm, to get top mileage out of a statistical program, it must be integrated into one overall master plan—along with some prescribed procedures for working the final numbers directly into the company's planning and operating budgets.

Another administrative problem implied in setting up an integrated system involves coordination of all quantitative analysis in a multidivision company. Take the case of a firm that originally had each of its nine operating divisions reporting directly to corporate headquarters. This was supposed to give top management an overview of current and projected operations. Unfortunately, each division had its own views on how to prepare reports. There was little indication of which divisions tended toward conservatism, toward overoptimism, or, indeed, toward general accuracy.

After having been burned a few times, the company decided to change all this, spelling out the ground rules for statistical analysis and projection. The new rules weren't so rigid as to preclude the use of differing techniques by each division. But the divisions were required to adhere to certain standards of accuracy—and to justify in writing when and why they were using an unrecommended approach.

A closely allied problem faced by many firms is the need to set up detailed instructions for reconciling findings or recommendations worked up by individual divisions and those worked up by corporate headquarters. Sometimes there's a logical reason for divergence—each echelon may view the picture from different perspective. But in any case, a system of checks and balances must be set up to ensure the most reasonable course of action consistent with company goals.

Many large, diversified manufacturers tackle the subgroup-coordination problem through periodic meetings. First, the individual division prepares its own version of, say, a sales forecast under some very broad, general guidelines handed down by top management. Then when the forecast is ready, division personnel sit down with corporate personnel to iron out any differences of opinion.

The audit is still another popular way of coordinating statistical reports. Every divisional or regional study is funneled through a corporate group. This group, made up of both marketing and statistical experts, goes over the individual reports to see whether they conform to both the standards and the general marketing policy set up by top management.

Any auditing system worth its salt should also include provision for follow-up on all original recommendations and findings. Equally if not more important are the steps taken to see that errors, if they occur, are caught quickly.

Also essential is the need to set up procedures for assuring that the next report will be at least as good as—if not better than—the last one. Postmortems can be of considerable help in this area. They stimulate thinking, point up where previous reports went off the track, and generally contribute to continuing improvement in quantitative analysis.

In any drive toward improvement some key questions to ask would include: (1) Can current techniques be improved upon? (2) Are the assumptions realistic and checked on from time to time? and (3) Is the company getting top mileage out of the current effort? On the latter point, many firms beam their market intelligence to only selected areas of management—completely ignoring others that might gain marginal benefit from such reports.

Presentation Problems

The way in which quantitative results are presented to top management can also play a crucial role in the success or lack of success in a corporate statistical program.

The key point to remember for those on the reporting end is that time is money for the people who will be acting on the recommendations or findings. Hence the emphasis must be on clear-cut functional presentations—without the scores of footnotes and qualifications that might interest a professional mathematician. In short, it's the "big picture"—not methodology or details—that needs to be gotten across.

If, however, some technological information is necessary, it is usually better to incorporate it into an appendix rather than into the body of the report. This assures that the supporting evidence won't interfere with the smooth flow of reading, and yet management can, if interested, delve a bit deeper into the mathematical procedures and basic anatomy of the report.

The way in which the report is delivered is another important consideration. It should be easy to understand and should be directed toward the businessman and phrased in his kind of language. In short, the presentation should be "alive"—and able to bring the basic message home with impact, clarity, and simplicity.

Sometimes an oral rather than a written presentation may be preferable. Or perhaps an oral presentation can be made first, with the detailed written report following. The advantage of the oral approach is that it permits the use of visual aids, giving the report the added advantages of surprise and showmanship.

Charts should be used liberally whether the approach is oral or written—with a lot of color and overlays if possible. When feasible, these should replace tabular material, which can be deadly when overused. In the words of the old saying, "One picture is worth a thousand words."

But while trying to make the report palatable, don't go overboard the other way and make it nothing but a set of glittering generalities. If management is to act, facts have to be presented. This also means avoiding excessive hedges, which make findings virtually worthless.

On the other hand, the basic limitations of any study must also be pointed out—specifically, (1) the key assumptions on which the findings are based, (2) the underlying approach used to arrive at the results (market research, correlation, simple extrapolation of past events, etc.), and (3) the probable range of error.

On the latter score it should be made amply clear that the data and recommendations presented are the most probable in the light of the available market intelligence. Avoid giving the impression that they are 100 percent sure.

In the case of a forecast, the range of error should be spelled out in detail. Thus if sales are expected to rise by, say, $1 million in the following year, point out how much this figure could vary, on the basis of the laws of probability. Telling what the figures can and cannot do protects both the originator and the recipient of the report and prevents some ticklish misunderstandings from developing later on.

TWO

The Use and Misuse of Averages

An average means many things to many people. To some it's a handy way of quickly describing a mass of data. To others it's the first step in a series of analytic statistical yardsticks. And to still others it's a convenient way of sweeping things under the rug.

Even to the statistician the average is a multidimensional measure. That's because the numerical result can differ significantly—depending upon which of a half dozen or so concepts of the average is employed.

In no way is this meant to denigrate the concept of the average. It's useful, it's needed, and it fills a recognized purpose in the business world. At the same time, it should be kept in mind that averages can sometimes be misleading and subject to considerable abuse.

But on one point there can be very little disagreement: Averages are popular. This popularity probably stems from a combination of several factors, including:

1. Simplicity: The underlying concept of an average—that of a measure of central tendency—is understood and accepted by all.

2. Ease of computation: The arithmetic involved is painless.

3. Usefulness as a policy aid: By putting the spotlight on the common denominator of the data rather than, say, on their variation, averages make decision and rule making a little easier.

4. Usefulness as a building block: An average is almost always a necessary first step to more sophisticated statistical analysis.

5. Opportunities for distortion: Many things can be hidden as well as revealed by an average.

Each of these five points is treated separately below.

1. *The concept of central tendency can be easily grasped—even by people untrained in the mathematical disciplines.* Thus it is almost natural, when asked to describe any mass of data, to answer in terms of the average. Ask a person about the income of a doctor, lawyer, or Indian chief, and the average is the first quantitative measure that ordinarily comes to his mind.

This is due to a large extent to the inherent nature of data. Whether the data concern income, the width of a bolt, or the number of sales per day, the observations will tend to vary within a certain range. What's needed is a simple figure which can fairly and accurately represent all the observations. Such a figure is the average.

In large part, emphasis on the average is also psychologically based. In almost all of us there is a need to find a common denominator when describing any particular phenomenon. The human mind abhors disorder or complex descriptions, so what could be better than to simplify by describing the typical or average? It's neat, it's precise, and it does away with untidy qualifying statements.

In the business world it often provides a "quick and dirty" handle to a large mass of seemingly unwieldy data. Thus knowing the average income or age of people buying a certain product can at least give the businessman a rough first approach to his potential market. It is a starting point—and one has to start somewhere.

The only trouble, as explained below, is that the starting point is often the stopping point. Because of lack of knowledge, lack of money, lack of time—or just plain laziness—too many are prone to take the average as the ultimate in statistical revelation.

Unfortunately, business just isn't amenable to this kind of simplistic approach. To be sure, averages serve a useful purpose. But they are basically a limited tool for describing complex quantitative data. Much more is needed than just a measure of central tendency to evaluate and analyze the flood of data that comes across the average businessman's desk.

2. *The computational procedures for calculating the various measures of central tendency are relatively simple.* One doesn't have to be a graduate economist or mathematician to work up a simple average.

Thus, in the case of the most popular average—the arithmetic mean—all that is required is knowledge of addition and division. And even for some of the more esoteric measures of central tendency, such as the geometric or harmonic mean, the formulas and computations are for the most part on the unsophisticated side. Certainly, they are within the grasp of any individual who has normal intelligence and a little formal training in mathematical disciplines.

This ease of computation is in many ways an incentive even where it is recognized that more complex statistical measures might be more useful. Thus a firm that wants to know something about the purchasing power in its marketing area can obtain the average with relative ease and at little expense. If, however, it wants to know something about the range of incomes or breakdown by ethnic group or education, the work becomes a full-sized project—requiring research, effort, time, and money. Ergo, the choice is often to take the path of least resistance and hope for the best.

3. *Simplifications wrought by use of an average can often prove extremely valuable in the formulation of general policy or rule-of-thumb procedures.* A case involving the entire American economy comes to mind.

Specifically, it is now generally recognized that wages, productivity, and prices are closely tied together. As long as productivity gains keep up with hourly pay boosts, there is little wage cost pressure to boost prices.

With this in mind, economists a few years ago derived the concept of wage-price guideposts for United States businessmen and unions to follow. Essentially, it stated that all wage increases should be kept in line with the average productivity gain for the entire country.

The guidelines, of course, recognized that not all industries and all firms have productivity advances that conform with the overall average.

So a further refinement was made stating that if wage boosts suggested by the guideline approach should exceed a company's productivity gain, some price increases were permissible. On the other hand, if wage boosts fell short of productivity advances, prices should be lowered accordingly.

Thus the wage-price guideline concept didn't shut out the possibility of upward and downward movements in prices. But rather, by hooking everything into a national average productivity figure, it was hoped that the average prices would also tend to remain stable.

And that's exactly what happened over the period 1963 to 1966. After that, however, the concept broke down because it was based on voluntary rather than mandatory compliance. Nevertheless, attempts are still being made to revive this valuable concept—primarily because experience shows that the averaging idea behind it is sound.

Let us come down now to a more mundane type of business problem. Some firms require little or no expense account documentation as long as outlays are in line with a firm's average experience. They will OK a higher-than-average amount—but will generally require more extensive documentation because it varies significantly from the average.

4. Averages are almost always needed as a general first step toward, or building block for, more sophisticated approaches.   And since it must be computed, the average is presented as a piece of statistical information or intelligence—even though its contribution to overall understanding of the problem at hand may be minimal.

Thus almost any problem where the variation in the data is to be studied would require the calculation of the arithmetic mean. And with good reason. For a study of variation must answer the question, Variation from what? And what could be better in this case than variation from the yardstick of central tendency—our old friend the average?

Averages also play key roles in the computations needed for correlation, trend analysis, model building, linear programming, etc. Put another way, the average is the cornerstone or base of any complex statistical program. Remove it or ignore it, and the whole edifice is in danger of collapse.

Some Common Distortions and Errors

The fact that averages can often sweep a lot of unwanted information under the rug is also often an incentive for its use. The economist who talks of the extremely low unemployment figure for adult men is usually trying to impress his listeners with the general high level of prosperity and well-being in the country.

This may hold true for the nation as a whole, but it most assuredly doesn't hold true for all the segments of the population that make up this national average. Thus a closer look at the statistics would show that unemployment is still a serious problem among minority groups and among young unmarrieds. And if you further subdivide into unmarried minority groups, the figure rises to disturbingly high levels.

In short, to say that no unemployment problem exists because the overall jobless average is low is to present an extremely misleading picture. And if it is used as the sole basis for policy decisions, the inequities and pockets of poverty that still plague our nation will escape needed scrutiny.

Price indices (averages of many different prices) can lead to similar misunderstandings. Consider the popular cost-of-living index. During 1968 this widely used inflation yardstick rose by more than 4 percent—leading many to believe that manufacturers and retailers were either passing along sky-high wage increases or simply fleecing the public.

Neither view, however, was justified. For a breakdown of the index quickly revealed that the price rise was centered mostly in services, which rose by close to 6 percent, and food, which (because of a poor crop year) went up by close to 4 percent. On the other hand, those items under direct control of big labor and big management (autos, appliances, etc.) rose by only 1 percent.

Again, a closer look at the component parts of an overall average would have given the reader a better understanding of what was happening—one which might then suggest proper programs and action for correcting a very unpleasant situation.

A somewhat different illustration of how averages can mislead cropped up a few years ago when antipollution programs first began to present a problem to industrial management.

During that early stage it was found that United States industry was spending hundreds of thousands of dollars on control equipment that was undersized for the jobs that had to be done to meet applicable regulations on air and water purity. And the principal reason for this was that sellers were specifying and buyers asking for equipment that would be adequate for average conditions—but no more.

Take the problem of stack emissions of sulfur dioxide. These may vary from a low of 5 parts per million all the way to a high of 5,000 parts per million. An average figure obtained after taking stack samples over a two-month period would be meaningless because the permissible limit could still be greatly exceeded much of the time.

Averages can also be confusing because of the wide range of different types available to the business analyst. Thus the arithmetic mean, the median, and the mode can all be used. All yield different answers, and yet all can be valid under certain circumstances—as long as it is properly understood what is being measured. It is a different underlying concept of what constitutes central tendency that leads to the different results. This will become clearer when these and other averages are discussed below in greater detail.

At times the choice of a particular average can be critical. More to the point, there are instances where choice of one over another can lead to completely erroneous results. Sometimes this is done deliberately, as when one quotes an arithmetic mean to hide, say, a distorted income distribution. In other cases where special types of problems are involved, more esoteric measures of central tendency such as the harmonic and geometric mean may be the only appropriate ones.

In any case, it should be clear by now that the average isn't exactly a simple, all-embracing, all-useful panacea for solving today's complex business and economic problems.

However, before going into some of the more sophisticated statistical techniques, it might be useful to go over each of the more commonly used measures of central tendency—pointing out their advantages, disadvantages, and pitfalls, as well as how they can be applied in solving some of today's more common business problems.

It should also be realized that the choice of an average itself presents a problem—specifically, deciding which measure should be used to best describe a given set of observations.

This decision will, of course, depend on the advantages and disadvantages of each measure (these are detailed below). But in addition, the final decision may depend on the nature of the distribution of the data in question. Thus if there are extreme deviations on either the upside or the downside, the arithmetic mean may not give the most meaningful picture of central tendency.

Another factor in deciding upon a given measure is the concept of central tendency desired for the particular purpose. Thus if a typical family is needed to determine marketing policy, the median or the mode may be the best average to use.

Sometimes it may even be expedient and useful to calculate two or three different measures. If properly explained and documented, this can reveal a lot more about the basic data than any one single measure. All this will become clear as the specific measures are analyzed below.

The Arithmetic Mean

In its simplest form this measure is defined as the sum of all the observations in a given set of data divided by the number of observations. The important thing here is that it is the size of each and every observation (how high or how low) that ultimately determines the average value.

This is in contrast to some of the other measures of central tendency (such as the median and the mode), where it is the relative position of the observations (rather than their value or size) that determines the magnitude of the average.

To illustrate: The arithmetic mean of the three numbers 1, 2, and 6 is obviously 3. But the median or value of the middle item is 2. Thus these two measures of central tendency yield significantly different values.

Despite differing results, the arithmetic mean is generally the most popular and most useful of the averages—and for good reason. The following are some of its big pluses.

1. *It facilitates understanding.* This has been touched upon earlier. For most people the average is synonymous with the arithmetic mean. In fact, many are unaware that other measures of central tendency even exist.

2. *It eases the chore.* There is no need to array the data, as in the case of a position-oriented average. And make no mistake about it—arraying can be a backbreaking job when tens of thousands of individual observations are involved.

3. *It reduces sampling variation.* It can be shown through higher mathematics that the arithmetic mean of, say, 50 sample observations is likely to be a better approximation of the underlying average than, say, a sample median based on the same 50 observations. This is an important consideration, since sampling is a must in most types of statistical analysis in the business field.

4. *It facilitates algebraic treatment.* This is particularly important in sampling, where arithmetic manipulation of the data is a virtual

necessity. This mathematical plus also helps make the arithmetic mean the basic building block for more complex statistical measures. Most other averages are simply not suited for further treatment.

Closely allied with this advantage is the ability to combine averages of subgroups to find a grand or overall average. Only the value of each subgroup mean and the number of observations in each subgroup are needed. Thus if the average (arithmetic mean) labor cost is known for individual operating divisions, it is a relatively easy task to calculate the average labor cost for the entire firm.

On the other hand, when working with subgroup medians or modes, the entire array of observations is needed to work up an overall average.

Another plus stemming from the arithmetic mean's algebraic maneuverability is that it can be shown that the sum of the deviations about the arithmetic mean is zero and that the sum of the squares of the deviations is a minimum. These particular properties make the arithmetic mean ideally suited as a frame of reference for measuring a series' inherent variability.

But the arithmetic mean isn't without some disadvantages. It is clear, for example, that because the size of each observation plays a part in the final result, the measure can be distorted by extreme values. This is not true of the other popular measures of central tendency.

A case in point: the averaging of income for all individuals in a Latin-American country. Because there are some extremely rich individuals in a land of many poor, averaging via the arithmetic mean could be misleading. It would tend to raise the average because one rich man earning, say $1 million would carry the same weight in the average as 1,000 poor families earning $1,000 each. In other words, the one rich man would pull the average way up.

Clearly, this could be misleading to an analyst trying to judge the market potential for his product. Ditto for an economist trying to judge the nation's standard of living. Averages of position eliminate this distortion. Thus the middle income in this Latin-American illustration would be somewhere in the poverty range—a much more realistic appraisal of the situation.

In some few instances use of the arithmetic mean is simply not appropriate—leading to a wrong rather than just a distorted answer. The classical examples here revolve around data presented in ratio form—dollars per day, miles per hour, etc. Depending upon how the problem is phrased, either the arithmetic mean or a variation of this measure, known as the *harmonic mean,* might be called for. More about this below.

At other times simple averaging—because it ignores weighting—can lead to some strange results. Consider a product with three basic costs—labor, raw materials, and overhead—all of which add up to $1 per unit. Of this, labor accounts for 50 percent of the total, or 50 cents; raw material, for 30 percent, or 30 cents; and overhead, for 20 percent, or 20 cents.

Next assume that labor costs double and that raw-material and overhead costs go up 50 percent each. A simple average (100 percent plus 50 percent plus 50 percent—all divided by 3) would lead one to believe that costs were going up 66⅔ percent.

But this kind of simplistic approach ignores the fact that the item that went up most (labor) is the cost that bulks heaviest. Ergo, the result understates the true increase.

A dollars-and-cents follow-through will illustrate the degree of understatement. The new labor cost is $1 per unit (50 cents doubled). The new raw-material cost is 45 cents per unit (a 50 percent boost in a 30-cent cost). The overhead charge rises to 30 cents (again a 50 percent boost, but this time applied to a 20-cent-per-unit item).

In short, the total new cost is $1.75 per unit ($1 plus 45 cents plus 30 cents)—or an increase in total costs of 75 percent rather than the 66⅔ percent hike suggested by the unweighted approach.

The same sort of problem crops up when the averaging of averages is involved. Take the case of a company that wants to ascertain the average hourly wage in two of its individual factories. The average wage in one factory, which employs 100 men, is $3.50 per hour, while the average in the other, which employs 200 men, is $4 per hour. A simple unweighted average would yield a value of $3.75 per hour—an underestimate because more men are being paid $4 per hour than are being paid $3.50 per hour.

The correct approach is to weight as follows:

$$\frac{(100 \times \$3.50 + 200 \times \$4)}{300 \text{ (total no. of workers in both factories)}} = \frac{\$350 + \$800}{300}$$

$$= \$3.83$$

In essence, what has been done is to assign twice as much weight to the second subaverage—since twice as many men are involved.

Finally, it should be noted that the arithmetic mean—indeed, any average—is virtually meaningless when the values are widely dispersed. If there are two houses on a street—one populated by a millionaire, the other by a relief recipient—it is meaningless to point out that the average income on the street is, say, $500,000 per year. In short, there should be a tendency to cluster for any average to be valid.

The problem of weighting will also be explored in Chapter 6, where index numbers are discussed. Index numbers are basically the arithmetic average of observations of varying importance.

Before leaving the subject of arithmetic means, it might also be useful to take a simple example and illustrate how different measures of central tendency more often than not yield different results.

Take a hypothetical case of the average width of steel plate coming off a production line. Specifications call for a 1-inch width—but because of the inherent variability in the production process, the width of individual production runs tends to cluster about rather than precisely hit the 1-inch target.

TABLE 2-1   *Solving a Problem*

Width	Frequency (no. of production runs)	Width × frequency
0.97	1	0.97
0.98	2	1.96
0.99	30	29.70
1.00	24	24.00
1.01	15	15.15
1.02	18	18.36
1.03	11	11.33
1.04	4	4.16
Total	105	105.63

$$\text{Arithmetic mean} = \frac{\text{width} \times \text{frequency}}{\text{no. of observations}} = \frac{105.63}{105} = 1.006$$

The table above illustrates the actual calculation of the arithmetic mean. This entails (1) multiplying the number of observations in each width category, (2) summing up, and (3) dividing by the number of observations. The result is an arithmetic means of 1.006—somewhat above that called for in the specifications.

It should be pointed out that the calculation presented here for determining the arithmetic mean is the simplest one. There are others which

(1) deal with more complex types of grouped data and (2) reduce the arithmetic by algebraic manipulation. But these involve use of formulas and symbols.

True, they are far from difficult. But their explicit descriptions are not deemed necessary here—principally because the aim of this book is to emphasize basic function rather than statistical formulations. Any beginning text will generally present a rundown of these formulas for the arithmetic mean—along with appropriate recommendations on when and where to use them.

Averages of Position

Using the same basic data, the median and mode (measures of central tendency based on position rather than value) yield different results in the majority of cases.

The *mode* is defined as the most frequently occurring observation. Under certain circumstances it can be considered the most typical or normal value. In the case of the steel plates, it would be simply the width with the highest frequency of occurrence—0.99.

The *median* is defined as the point at which a series divides into two equal parts—or, more simply, the value of the middle observation when all the observations are arranged according to size. In our steel-plate example there are 105 observations—so the task is to find the fifty-third, or middle, item and assign to this one the position and value of median.

The fifty-third observation is obviously buried in the 1.00 class. Hence the value of the median is 1.00.

Thus we now have three separate values for central tendency: a modal value of 0.99, a median value of 1.00, and an arithmetic mean of 1.006.

A graphic illustration can perhaps best present the results (see Figure 2-1). A first glance at the figure reveals that the distribution of steel-plate widths is asymmetrical—that is, it is off-center, leaning over toward the right. In the parlance of statisticians, the distribution is skewed to the right.

This is simply a reflection of the fact that there are more observations on the upper side of the modal or peak class than on the lower side. When this is the case, the three measures of central tendency always diverge—with the mode on the low end, the median in the middle, and the arithmetic mean coming up with the highest reading.

On the other hand, when a distribution is skewed to the left, the relationship between the three measures of central tendency is reversed. The arithmetic mean is now the lowest, the median remains in the middle, and the mode is the highest. The fact that the mean is always on one or the other extreme reflects the fact that this measure—as noted above—is unduly affected by extreme values.

By how much do these three measures of central tendency differ? The exact amount, of course, will depend on the amount of skewness or asymmetry inherent in the distribution. But statisticians have a rule of thumb for a moderately skewed distribution: The mode and the mean are farthest apart. Moreover, the distance from the mean to the median is one-third the distance from the median to the mode. For those who are mathematically inclined, this rule can be stated by the formula

Mode $\cong$ mean $-$ 3(mean $-$ median)
0.99 $\cong$ 1.006 $-$ 3(1.006 $-$ 1.00)
0.99 $\cong$ 1.006 $-$ 3(0.006)
0.99 $\cong$ 1.006 $-$ 0.018
0.99 $\cong$ 0.988

where $\cong$ stands for "approximately equal."

The three measures can coincide only when the distribution of ob-

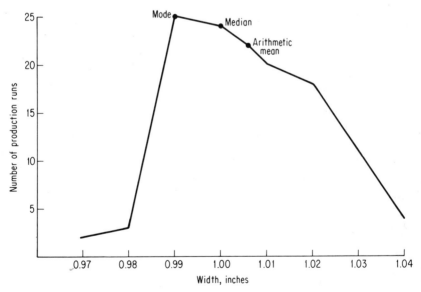

FIGURE 2-1   *Average width of steel plate.*

servation is perfectly symmetrical. But the chances of this occurring in the uncertain world of business are indeed slim—mainly because so many different variables influence the individual readings. Thus in the case of the steel-plate widths, the variation could be due to poor workmanship, varying qualities of raw material, machine malfunctioning, etc.

Before leaving position type of averages, a few words on their advantages and disadvantages may be in order.

The pluses, of course, stem from the fact that the median and the mode are not affected by extreme values. Thus in the case of the steel plates, one might be tempted to modify the process in view of the fact that the arithmetic mean was running above specifications. But a further look would reveal that this deviation was due in large part to a few very high readings—that most items (in this case 69, or more than two-thirds) fell within 0.01 of the target width. Looking at the arithmetic mean alone would not suggest this degree of closeness to the specified width.

Another advantage of the middle-position concept is that it can be subdivided into smaller categories. Indeed, just as the median cuts the distribution into 2 equal sections, so quartiles can cut it into 4 equal parts, deciles into 10 equal parts, and percentiles into 100 equal parts. In short, by focusing on position rather than value, it is possible to make a more detailed study of any distribution.

The most common use of these smaller subdivisions is in the field of education, where a student may be said to be, say, in the upper decile, or 10 percent, of his class. But there is nothing that disqualifies this type of analysis for use with business-oriented problems.

Suppose, for example, that we are interested in the price/earnings ratio of stocks. By using this approach, we could say, for example, that 1 percent, 10 percent, or 25 percent of all corporations have P/E ratios above a certain level. This, in turn, could be of considerable use to a stock market analyst appraising the relative merits of specific stocks.

The lack of algebraic maneuverability—touched upon above—is probably one of the key drawbacks of both the median and the mode. In addition to cutting down on their use as building blocks for more sophisticated approaches, this mathematical shortcoming seriously reduces their usefulness in sampling. Thus, if one mode is, say, larger than another, it is virtually impossible to determine whether the difference is due to chance or not. The arithmetic mean, on the other hand, has no such limitation.

Other disadvantages are that (1) these measures aren't nearly as familiar as the arithmetic mean and (2) arraying of the data is necessary. In the case of the mode, an additional problem often crops up where there is a bunching up of observations in two parts of the distribution. This is known as a *bimodal* distribution—and often presents the problem of which "hump" or concentration to label as the mode.

Sometimes bimodality is the result of chance—and then there's little that can be done except to cite two modal measures of central tendency. But at other times bimodality may be due to the fact that the original distribution is made up of two sets of nonhomogeneous data. In this case a separating out of the two subgroups—each of which has a single mode—can solve the problem.

For example, the steel-plate illustration might have been derived from steel coming off two different machines. If this were the cause of bimodality, then a splitting out of observations for each machine might reveal that each hump was due to a specific machine, rather than to the inherent variability in the basic manufacturing process.

Other Measures of Central Tendency

While the arithmetic mean is by far the most important and popular yardstick—with the median and the mode ranking second and third—there are also other, more specialized averages. While classified by most statisticians as minor, they are nevertheless crucial for some types of calculations. Two such measures will be discussed here: the geometric mean and the harmonic mean.

Geometric mean.   This average is needed when the data are presented in terms of rates of change rather than in absolute numbers. Indeed, averaging via the arithmetic mean under these conditions would actually yield incorrect results.

Consider the simple example of a firm that quadruples sales in one year and then doubles them the next. A simple arithmetic mean would suggest that sales on the average tripled (the arithmetic mean of a quadrupling and doubling). Assume for a moment that the firm started out with $1 million in sales. An average tripling in each of the two years would then imply sales of $3 million after one year and of $9 million from a similar tripling the second.

But consider what actually happened. A quadrupling the first year meant simply that sales rose from $1 million to $4 million. A subsequent doubling would bring sales up to $8 million during the second year.

Clearly there's an inconsistency. The sales actually rose to $8 million, yet the use of the arithmetic mean suggests an average tripling to $9 million. In short, the arithmetic mean has in this case overshot the mark by a significant $1 million. Indeed, it can be shown that the arithmetic mean will always overestimate the true average when rates of change are involved.

What is needed, then, is a measure that will give the true average rate of change. The geometric mean is just that average. It is defined as the nth root of a product of n observations. More specifically,

$$\text{Geometric mean} = \sqrt[n]{X_1 \cdot X_2 \cdot X_3 \cdots X_n}$$

where n = number of observations

x = value of any given observation

Next let us apply the formula to the problem at hand:

$$\text{Geometric mean} = \sqrt[2]{4 \times 2} = \sqrt{8} = 2.83$$

Finally, as a check, use this average growth rate for each of the two years. The end result will be very close to $8 million—the actual sales result after two years.

The geometric principal is also involved when the problem is to assess average growth over a period of years and only the beginning and end years are known. For example, if sales over a two-year period doubled, our first inclination would be to say that sales rose 50 percent in each of the two years. Again this is an arithmetic mean approach, and again it would tend to overestimate the true growth rate.

A closer examination will prove the point. If the starting point is sales of $1 million, a 50 percent growth rate implies that during the first year they would rise to $1.5 million and during the second another 50 percent over the $1.5-million figure, to $2.25 million.

But the fact is that sales were only $2 billion after two years. So clearly the true growth rate was something under the 50 percent figure suggested by the use of the arithmetic mean. How much under? Again a little algebra and the use of the geometric mean will provide the correct answer.

First set up the formula for the geometric mean:

$$\text{Geometric mean} = \sqrt[2]{\frac{x_1}{x_0} \cdot \frac{x_2}{x_1}}$$

where x_0 = sales at the beginning period
x_1 = sales during the first year
x_2 = sales during the second year
Next simplify the formula:

$$\sqrt[2]{\frac{x_1}{x_0} \cdot \frac{x_2}{x_1}} = \sqrt[2]{\frac{x_2}{x_0}}$$

Next substitute the given values for sales at the beginning and sales at the end of two years:

$$\sqrt{\frac{\$2}{\$1}} = \sqrt{\$2} = 1.414$$

INTERPRETATION: Each year on the average is 1.414 the size of the previous year. Put another way, the average rate of growth is 41.4 percent.

This seems like a laborious procedure—and it is, particularly when a considerable number of years are involved. Happily, tables have been prepared which yield the growth rate without requiring any work. (They are nothing more than simple compound interest rate tables.) For more precise results, logarithms can be used to simplify the work.

This so-called compound interest formula to calculate the average rate of growth can also be applied to related growth problems. Note from the above formulation, for example, that we had as given data (1) beginning sales, (2) the number of years, and (3) final sales. The only unknown—the one we were solving for—was the average rate of growth over the period.

Mathematically speaking, three out of four variables were known. It follows, then, that knowledge of any three would be sufficient to determine the value of the fourth, or unknown, magnitude. Thus in addition to figuring out the average rate of change, this approach can be used to estimate (1) sales in some future year, assuming a constant rate of growth, and (2) the number of years it will take to achieve a given sales level, again assuming a constant rate of growth.

Rate of growth problems are often decisive factors in business decisions. Questions on sales-force expansion, capital spending, product development, etc., depend to a large extent on the average growth rate calculation. Hence the geometric mean is an important business planning tool.

Why does the geometric mean work while the arithmetic mean does not? It all stems from the basic fact that the geometric mean gives equal weight to equal rates of change, while the arithmetic mean gives equal weight only to equal absolute change. Thus the geometric mean counts a doubling from 2 to 4 as important as a doubling from 4 to 8—and this is what we are after when averaging rates of change. On the other hand, the arithmetic mean gives the jump from 4 to 8 a bigger weight and hence leads to overestimation.

But working with the geometric approach is not without its problems: (1) It is not as widely known as the other averages, (2) the concept behind it is a bit more difficult to grasp, and (3) technical considerations prevent its use when there are negative values or where one of the items is equal to zero.

Harmonic mean. This is also a mean pressed into service in some cases where the arithmetic mean gives incorrect results. Generally speaking, it is used when data are expressed in ratio form (cents per pound, number per dollar, etc.).

Thus we now have two specialized means: the geometric mean for rates of change and the harmonic mean for ratios. But there is one big difference. The geometric mean must always be used for averaging rates of change. The harmonic mean, on the other hand, is only sometimes applicable for the averaging of ratios.

But before enumerating rules on when and when not to use the harmonic mean, it might be a good idea to look at a classical case where the arithmetic mean yields the wrong results. The problem involves a man traveling from one point to another at 20 miles per hour—and then traveling back to his original starting point at 60 miles per hour. The question: What was his average speed during the round trip?

Again the inclination is to use the arithmetic mean—the averaging of 20 miles per hour and 60 miles per hour. The answer, of course, would then be an average speed of 40 miles per hour.

But after a little thought, it should become clear that this is not correct. Specifically, on the first part of the journey it took him three minutes to go a mile, while on the way back it took one minute to go a mile. The average mile, then, obviously took two minutes, which is the equivalent of 30 miles per hour—not the 40-mile-per-hour figure that the arithmetic mean seems to suggest.

The harmonic mean, on the other hand, will yield the true 30-mile-per-hour average. It is defined as the reciprocal of the arithmetic mean of the reciprocals of the individual observations. A mouthful, certainly.

But let's see how it works using the problem at hand:

$$\text{Harmonic mean} = \frac{1}{(\frac{1}{20} + \frac{1}{60})/2} = \frac{1}{\frac{1}{40} + \frac{1}{120}}$$

$$= \frac{1}{\frac{3}{120} + \frac{1}{120}} = \frac{1}{\frac{4}{120}} = \frac{1}{\frac{1}{30}} = 30$$

Now take a problem more related to business. Assume that widgets sell at the rate of three for $1 in the West and two for $1 in the East. Using the arithmetic mean to determine the average price would, of course, yield a figure of $2\frac{1}{2}$ for $1—or 40 cents per unit.

But again some further thought is required. In the East the unit price is $33\frac{1}{3}$ cents, while in the West it is 50 cents. Using this approach, the average would seem to be $41\frac{2}{3}$ cents per unit.

Both can't be correct—or can they? If we assume that the same number of widgets are sold in the East and the West, then clearly the $41\frac{2}{3}$ cents per unit price is the correct one. Use of the harmonic mean would yield this result.

On the other hand, if we assume that the same dollar amount is sold in the East and the West, then 40 cents per unit (the average determined by the arithmetic mean) is the correct figure. That's because three units at $33\frac{1}{3}$ cents and two units at 50 cents yield an average price of 40 cents per unit.

If we want to keep the dollar amounts the same in each of the ratios being averaged—that is, the denominators in our number-per-dollar ratios—then the arithmetic mean is called for. If we want to keep the number of units the same in each of the ratios being averaged—that is, the numerators in our number-per-dollar ratios—then the harmonic mean is called for.

And therein lies the rule of thumb for when and when not to use the harmonic mean: When all the ratios being averaged have the same value in the denominator, the arithmetic mean should be used. When all the ratios being averaged have the same value in the numerator, the harmonic mean must get the nod.

Incidentally, the use of this rule would have steered us to the appropriate mean in the average-speed problem noted above. Recall that the ratio involved was miles per hour. Since the number of miles traversed (numerators) is the same for both figures to be averaged, it follows that the harmonic mean would be the correct one.

The harmonic mean is sometimes also used in index-number averaging. By combining the arithmetic mean, which has certain biases, with the harmonic mean, which is subject to biases that vary inversely with those of the arithmetic mean, a mutual offsetting of biases is possible.

The Complicating Factor of Variation

The concept of variability is closely linked with that of averages. Clearly, two groups of data can have the same basic average, yet have significantly different degrees of dispersion. The average of 8, 9, and 10 is 9. So is the average of 5, 9, and 13. Yet the policy decisions suggested by the first array of data could differ sharply from those suggested by the second. Put another way, knowing the average isn't nearly enough.

What it all boils down to is the fact that the average per se is a very limited tool unless the variation about that average is known. If it is so great that there is very little central tendency, then the average is of doubtful significance. Generally speaking, the smaller the degree of variation, the more meaningful the average.

But whatever the amount of variation, it can usually be traced to one of two basic causes: lack of homogeneity in the data or inherent variability. Take the first case. If wages of skilled and unskilled workers are lumped together, the spread will be larger than if the more homogeneous subgroups of skilled and nonskilled workers are considered separately.

Inherent variability is another problem. Little can be done about it—but that's not to say that it can't be used as a powerful analytic tool for decision making. In many instances, for example, the amount of such variability can aid in choosing between alternative procedures or processes.

Thus a manufacturer of light bulbs may come up with a system that yields a higher average life but greater variability. He may decide against it because uniformly high quality of the bulbs may provide a better selling basis than longer life with a substantial chance of early blowouts. When it comes to items involving safety, such as tires, it is clear that variability must often outweigh long life as the primary consideration.

The problem of measuring dispersion or variation is treated in almost any standard statistical text. Curiously enough, nearly all such measures involve the calculation of an average measure of variation. So again the concept of average is of crucial importance in statistical analysis.

By definition, then, a series with high average variation would clearly have a wider spread than one with low average variation. There's no problem about that. Difficulties, however, arise in the computational area. As noted earlier, the deviations about the arithmetic mean are zero. So any measure that would average deviations about the mean would always turn out to be zero—the pluses always canceling out the minuses.

In the case of a distribution of wide variation, big pluses would offset big minuses. Where small variation is involved, small pluses would just as efficiently cancel out small minuses.

In short, the fact that minuses and pluses are involved would seem to preclude a measure of average variation about a measure of central tendency.

But statisticians are an imaginative lot. So to eliminate the plus-and-minus problem, they have come up with a little-known average that goes under the name of the *quadratic mean*. This essentially sums up the squares of all deviations, divides this sum by the number of observations, and then finally takes the square root of the result to get back into the original units.

Application of the quadratic mean approach in the variation problem immediately eliminates the problem of averaging minus and plus deviations—for the minus figures multiplied by themselves by definition result in positive figures. The resulting measure of variation is commonly called the *standard deviation*. Aside from providing additional information about a distribution, it is sometimes of crucial importance.

This is certainly true in the case of statistical quality control, where a given level of variation is used as a yardstick for acceptance or rejection of a given lot. Specifically, acceptance and rejection regions are set up, on the basis of the standard deviation. Then those lots which fall beyond the rejection limits are deemed to be unacceptable.

Equally important—variation, or more specifically the standard deviation, is the basic block for even more sophisticated statistical analysis. Thus correlation, where variation in one series is statistically related to variation in another series, is inherently based on the concept of the standard deviation. Indeed, one way to explain correlation is in terms of explained and unexplained variation.

There is another important use of the standard deviation—the estimation of how many observations might fall between any two values in a given distribution. It can be shown, for example, that if the distribution is fairly symmetrical, or normal, about two-thirds of all observations will fall between plus or minus one standard deviation of the mean. Plus or minus two standard deviations would cover 95 percent of the cases, and plus or minus three standard deviations would cover virtually every case (see Figure 2-2).

A practical example might involve a maker of clothing, to whom sizes are a prime consideration. After calculating the average and the standard deviation, he might manufacture sizes ranging only to two standard deviations on either side of the average. This would assure him of reaching approximately 95 percent of his potential market.

So far, what has been discussed is absolute dispersion. But often what is required is a measure of relative dispersion—particularly when one wants to compare variation in two or more series. Generally speaking, two such series cannot be compared for absolute variation when (1) the means are significantly different or (2) they are expressed in different units of measurement.

But there's a way out. Again our old friend the average comes to the rescue. By dividing the respective standard deviations by their arithmetic means, both standard deviations are put on a comparable basis.

Clearly, a standard deviation of $2 in a share of stock that averages $5 per share does not have the same intensity of variation as a standard deviation of $2 in a stock that averages nearly $100 per share. Common sense would tell us that the stock averaging $100 per share and varying by $2 has less inherent spread.

Divide both standard deviations by their respective means, and this becomes clear. In the case of the $100 stock, the relative variation is 0.02 (2/100). In the case of the $5 stock, the relative variation is 0.4

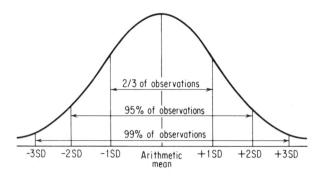

FIGURE 2-2   *Normal distribution of data.*

(2/5). Thus the $5 stock is 20 times as variable. As usual, statisticians have a name for this approach: the *coefficient of variation*.

A somewhat related problem arises when it is necessary to compare variations of individual observations about averages. Assume, for example, that employment aptitude in a particular firm is determined on the basis of abstract thinking and manual dexterity tests. A prospective employee gets 150 on the abstract thinking test, where the average is 130 and the standard deviation is 10. He gets 100 on the manual dexterity test, where the average is 85 and the standard deviation is 7.5.

The question then arises about this prospective employee's relative standing in both these crucial areas. This can be obtained by (1) taking the deviation from the mean for each test and (2) dividing the deviation by the appropriate standard deviation. Thus:

$$\frac{150 - 130}{10} = \frac{20}{10} = 2 \qquad \text{(abstract thinking)}$$

$$\frac{100 - 85}{7.5} = \frac{15}{7.5} = 2 \qquad \text{(manual dexterity)}$$

In this case it is apparent that the relative standing of the prospective employee in both cases is the same: two standard deviations above the mean. This approach is often referred to as a *standard score*.

One last note on variation: The standard deviation in its varying forms and uses discussed above isn't the only measure of dispersion available. As in the case of averages, there are alternative procedures. But also as in the case of averages, there is a most useful one—and that one is the standard deviation.

THREE

The Art of Charting

"One picture is worth a thousand words." An old saying—yes. But it's particularly appropriate in today's hectic business world. With executive time in short supply, pictures or graphic presentations are a virtual must—summarizing information almost instantaneously and presenting ideas more quickly and clearly.

There's an added impact advantage to the graphic approach. A well-designed chart or series of charts—in addition to transmitting a message more quickly and efficiently than any other form of communication—also tends to make a more lasting impression. Thus a chart showing sales doubling or being halved is a lot less likely to be forgotten than a dry statistical table revealing the same basic information.

Further, such illustrations as charts, diagrams, maps, etc., can clarify complex or hazy points. Put another way, the graphic approach in many cases is admirably suited to communicate facts and concepts much more efficiently than either words or figures. Thus the *break-even* point—a complex relationship between costs and revenue at various levels of operation—can be quickly explained by the use of the relatively simple and popular break-even chart (see page 68).

Charts also help pinpoint relationships that might otherwise be over-looked when presented in tabular form. If spending is plotted directly against income, for example, the correlation between these two key vari-ables becomes immediately apparent.

A closely related plus is the fact that charting facilitates comparisons much more easily than a column of numbers. Plot sales of new products against sales of relatively established products, and the growth potential of the new products may become a lot clearer. In the same vein, the use of charts can facilitate the comparisons of, say, sales growth patterns in one area of the nation with those in any other area.

In short, skillfully designed charts can help close the communications gap by presenting information vividly, concisely, dramatically, and in an easily understood way. More to the point—charting can provide businessmen with a powerful tool to get across the facts behind the figures and to increase the effectiveness of both written and oral communication.

Moreover, with today's information explosion, the graphic technique is admirably suited for any one of a thousand areas where simplified business reports and summaries are required. Thus with the aid of charts, a nontechnical executive may be able to grasp the significance of a new research and development proposal, or the production vice-president may understand the myriad of implications behind financial statements.

But like all statistical disciplines, charting and graphics are not without their pitfalls. If a chart is cluttered or confused, it can detract from a report rather than add to it. Similarly, outright mistakes—either pur-posely or inadvertently made—can hide or distort your message rather than clarify it.

Certain limitations of the approach should also be recognized. Charts cannot show quite as many facts as tables. So backup tables should be made available for those who want to delve more deeply in the information. Closely related is the fact that charts give only the "big picture"—and generally permit the reading off of only approximate values. Again this may call for tabular backup to permit greater detail. But these are minor disadvantages—more than offset by the added effectiveness a chart possesses in comparison with, say, a tabular presentation.

Recognition of the role charts can play, however, is only the first step—albeit an important one—in effective graphic communication. Other more substantive action is also required. Thus any overall strategy for visual presentation should include:

▪ Charting ground rules which outline the general policy on do's and don'ts—including special considerations applicable to your own firm's particular problems

▪ A thorough understanding of the basic statistical rules involved in the setting up of charts and graphs

▪ A detailed inventory on the type of charts and graphs that are now available for business usage

▪ A closer look at some of the ways in which charts can distort the basic message

Each of the above points will now be discussed in more detail.

The Charting Ground Rules

Graphic presentation is essentially a tool. And like any tool, it can have a payoff varying from excellent or good all the way to downright poor—depending upon how well it is used. But use involves more than just the specific choice of charts. An important prerequisite is the setting up of procedural ground rules, for without such overall direction, charting becomes a hit-or-miss proposition—losing much of its potential to both save time and increase comprehension.

It would be well to consider the following as part of such basic strategy.

1. *Keep the chart's textual material to a minimum.* If a chart requires a page of footnotes to explain its meaning, the approach loses much of its visual impact. Thus one-line titles with a virtually self-explanatory message are the kind that pay off. Indeed, even if the chart should require further explanation, it is better to oversimplify.

For example, a chart on copper consumption in the United States might be based on 85 percent direct coverage, with the additional 15 percent based on complex statistical estimating procedures. Despite these problems, the chart should be labeled "United States Copper Consumption"—for that is the basic message. The estimating qualifications are for the technically oriented rather than the harried business executive—and either these should be explained in the written report or else the report should direct the interested reader to places where he may find the additional computational details.

2. *Key charts to the salient points of the report.* The chart should illustrate the main thrust of your message. If it doesn't, it may lead to more confusion than light. Thus, if you are presenting a report on sales, the preponderance of graphic material should be on this subject and not on allied subjects such as material and labor costs or profit derived from these sales.

On the other hand, avoid too many charts which elaborate on the main theme. If the message is refrigerator sales, avoid the temptation to use graphs for every single model—even though the data may be readily available. Too many charts can be almost as enervating as a straight textual presentation. In short, too much of a good thing can be deadly.

3. *Work closely with the charting technicians.* The actual illustrator is an important cog in the process. He can often visualize whether the chart you suggest will have the intended impact. Equally significant, if the illustrator understands what the chart is trying to get across, he can often come up with suggestions for highlighting the message.

Close liaison between the charting technician and the people responsible for delivering the report can also prevent costly or time-consuming mistakes. Experience has shown that the chartist—if left alone with the original instructions—will often produce a product somewhat different from the one envisioned by the analyst. This can be due to either unclear directions or the fact that so much in the actual physical charting is subjective (choice of color, choice of scale, choice of print size, etc.).

4. *Allow for ample preparation time.* Too often people tend to come to the graphic arts department one or two days before the actual presentation time and expect an impactful, professional job.

This rush approach results in either excessive overtime work or the working up of a less effective chart. Note, for example, that impactful four-color charts—the kind that really deliver the message—often take upward of a week to prepare.

It is also a good idea to allow a few extra days for corrections, receipt of new information, or simply a change in your basic presentation strategy.

5. *Dry-run your material at least once if the presentation is oral.* Nothing so disconcerts an audience as a speaker who stumbles over a chart—unable to pinpoint its significance. Also a dress rehearsal—when all the visual material is available—may help point up areas where the chart-text combination can be improved upon. Finally, if a projectionist or other aide is needed, it may be a good idea to bring him into the rehearsal to assure against any foul-up in timing.

6. *Dress up charts and other graphic material as much as possible.* A dull chart isn't much of an improvement over a dull text. Remember that one of the key objectives of a chart is to focus the attention of the viewer on your basic message. Thus, aim for (*a*) large charts, (*b*) large lettering and thick lines, and (*c*) plenty of color. Sometimes the insertion of a photograph or two adds a dash of needed realism.

7. *Space your charts out logically.* No matter how artistic your visual material is, it can't provide a substitute for planning. In short, beware of the pitfalls of basing your written script on what is illustratable. As a rule of thumb, charts should be prepared after your presentation has been organized—not before.

8. *Avoid unnecessary illustrations.* It is costly and time-consuming to include unnecessary illustrations, and generally they don't help you toward your goal of providing a clear and realistic picture. If your message can be delivered with one chart instead of two, then use only one. Above all, don't take up management's valuable time by showing cartoons and other gimmicks which are entertainment-oriented rather than information-oriented. Put another way—a few good charts with a good story line are better than many charts which are inserted essentially for padding purposes.

9. *Gear the size of the graph to the audience.* It should be large if displayed to a large group of people, small if shown to only a few. If the illustration is to be a slide, then its readability on a screen is an added consideration. As a general rule, the graph should be readily legible to the person farthest away from it who will be asked to read and interpret it.

10. *Make use of overlays, particularly in oral presentations.* This involves first presenting a simple chart, say, on sales, and then adding a transparent overlay sheet—usually in color—to point up some significant comparison with the sales line originally depicted.

Techniques of Chart Construction

Aside from the planning aspects noted above, there are some strictly mechanical and statistical considerations that must enter into any charting decision. These involve the following.

1. Simplicity. Don't try to show too many things on the same chart. A graph by its very nature is designed to present a "quick and dirty" picture of what is happening. If too many things are presented at the same time, the chart often becomes either unintelligible or, at best, difficult to interpret.

Often, for example, an attempt is made to show volatile price changes in each of a series of commodities. The lines intersect and zigzag across one another, so that the ultimate result is a hopeless tangle of lines. It is generally good policy to limit charts showing comparisons to only three or possibly four variables.

2. Documentation. Any graph—no matter what its shape, form, or content—must carry certain identifying information, including:

(a) TITLE: Aside from the "keep it simple" caveat discussed above, titles should generally be designed to give the "what, where, and when" of the information being plotted. Think of the title as being akin to the headline of a newspaper, which of course is aimed at getting the attention of the viewer and arousing his interest. Indeed, a large portion of the attraction value of the chart may well depend on the wording of the title.

(b) SOURCE: Generally indicated at the bottom of the graph, this is presented for two purposes: to lend credence to the message presented and to enable the viewer to obtain further information on the subject if he so desires.

(c) THE GRID: In most cases it is desirable to draw background lines on the chart. These serve as a visual aid in making comparisons and for reading off values from the chart. Generally, these should be limited in number—for too many such lines are more likely to confuse than facilitate evaluation.

(d) THE SCALE: The grid lines can be evaluated by means of a scale of values placed along both the horizontal and the vertical axes. The number of values enumerated usually coincides with the grid lines shown. Note, too, that the scale of values should be such that a major portion of the graph's overall area is used for plotting purposes. This, however, can lead to some distortion or misinterpretation (see "Distortions," below).

(e) SCALE CAPTIONS: Identification of what the scale stands for is, of course, necessary. Thus some inkling must be given as to whether the series being plotted are in terms of pounds, ounces, tons, dollars, cents, index numbers, etc. Surprisingly enough, even professionals sometimes forget this key bit of information—probably because they assume that everyone knows the units in which the chart is expressed. Such an assumption, however, is completely unwarranted.

3. Chart shape. The size and shape of the chart can be important. A long, skinny chart, other things being equal, will tend to exaggerate change. On the other hand, a flat, squat chart will tend to play down change. There are times when the need to attract attention may dictate the choice of one or the other. But for the most part, moderation is called for—for extremes run the risk of inviting misinterpretation. In a few instances (where the chart is part of a written report) size also may be limited by the amount of space available.

4. Graph paper. The analyst usually makes the first rough pass at the chart before submitting it to the illustrator for a finished piece of work. For this first approach ruled graph paper is an invaluable aid. Note that this ruled paper need not show equal space for equal unit changes (a simple arithmetic grid), even though this is by far the most common type. Special ruled paper is also available for ratios (equal space for equal percent changes). More will be said about these ratio papers in the section below on types of charts.

Selecting the Correct Chart

As pointed out above, the basic function of graphs is to explain and interpret complicated quantitative statistics and to present them in an easily understood form. In many cases this also involves making a choice—for the fertile minds of statisticians and mathematicians have managed to come up with an impressive array of graphic alternatives. And what may be best for one particular problem may not be so for another. Some of the more popular charts include:

1. The line chart. This is the simplest and by far the most popular of the graphic forms. More often than not, another line or two are added to yield a comparison with other statistical series. However, as noted above, anything more than three or four lines on any given line chart should be avoided because they tend to give the chart a cluttered appearance.

The principal advantages of a line diagram are that it is simple to construct, is readily understood, and is particularly well suited for problems where there are a considerable number of values to be plotted. In cases involving variation in a quantity at many different periods of time, for example, the line chart is usually the most effective way of transmitting the information.

There are many variations of the line chart (see Figure 3-1), and if several such diagrams are needed for a single report, it may be a good idea to take advantage of this choice to avoid monotony and hence possible loss of reader interest.

These variations include:

(a) THE SILHOUETTE CHART: Fluctuations from a norm or base or zero line are shaded to provide sharper contrast.

(b) THE HIGH-LOW CHART: The range of, say, a variable such as price over a given time period is connected by a vertical line.

(c) THE COMPONENT LINE CHART: Component parts of a total (each shaded differently) are plotted on top of one another to yield the trend in both components and total.

(d) THE PICTORIAL LINE CHART: A picture or a drawing is either superimposed on a background or added at various points along the line.

(e) THE CUMULATIVE LINE CHART: The basic data are cumulated over time or cumulated to include more and more subgroups.

It is possible both to cumulate up (from the lowest-valued subgroup up to the point where all subgroups are added in) and to cumulate down (from the total of all subgroups gradually down, so that only the lowest is included). These types of cumulative curves are commonly referred to as *ogives*.

The advantages of this cumulative type of approach can best be appreciated through a simple example involving a distribution of population by age groups. By cumulating up, it is possible to determine at a glance the number of people less than, say, twenty-five years of age. By cumulating down, it is possible to quickly ascertain all people, say, over sixty-five years of age. Thus cumulating up is sometimes referred

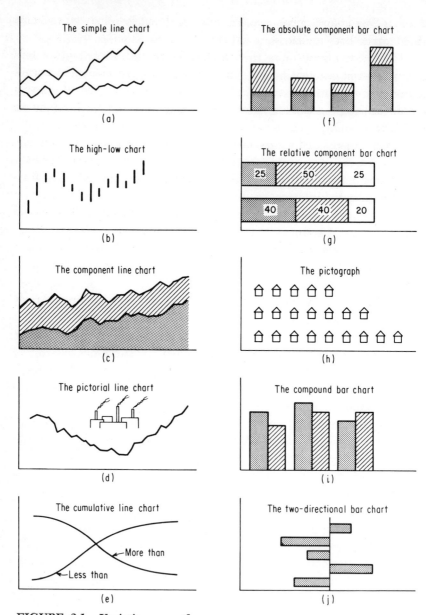

FIGURE 3-1 *Variations on a theme.*

to as a "less than" ogive, and cumulating down is known as a "more than" ogive. Both are shown in Figure 3-1.

2. *The bar chart.* This is probably the second most popular type of graph—and one that is also quite familiar to most people. In essence it involves contrasting quantities by a comparison of bars of varying length or height. The bars can be presented either horizontally or vertically.

Generally speaking, a bar chart will be more effective than a line chart when (*a*) the number of observations is few, (*b*) observations of different types or kinds are being plotted, or (*c*) observations from different geographic areas are being plotted. Sometimes when plotting a series over time, either a bar or a line diagram can be used. It should be kept in mind, however, that a line chart is usually better for interpreting the general trend over the entire period, while a bar chart is usually better for measuring specific changes from any one time period to the next.

As in the case of line charts, there are multiple variations on the basic theme. These include:

(a) THE ABSOLUTE-COMPONENT-PART BAR CHART. Component parts of a total bar (each shaded differently) are plotted on top of one another.

(b) THE RELATIVE-COMPONENT-PART BAR CHART. Each component is expressed as a percent of the total bar.

(c) THE PICTOGRAPH. A simple bar is replaced by a series of identical drawings or pictures, each of which represents a given magnitude. Hence the number of drawings or pictures on a given line is the equivalent of the length of a bar.

(d) THE DOUBLE OR COMPOUND BAR CHART. Two or more sets of data are compared over a period of time—with each set in a given time period separated from the other sets by a space.

(e) THE TWO-DIRECTIONAL BAR CHART. Bars are set in opposite directions to show positive or negative movement. Sometimes, to emphasize direction, the positive is shown in black, and the negative in red.

3. *The pie chart.* The so-called pie diagram is another commonly used device, particularly when share is the point that needs the major emphasis. The chart is essentially a circle or "pie," with the given segments of the circle representing different percentages of a total.

One common application is the "where your dollar goes" type of charts, which appear regularly whenever the federal budget is presented.

This same approach is also useful for the individual firm. Consider a case where company sales come to $100 million—with material costs accounting for $40 million; labor, $30 million; overhead, $10 million; and pretax profits, $20 million. A pie chart dramatically illustrates all this by splitting up a pie area and allocating 40 percent to material, 30 percent to labor, 10 percent to overhead, and 20 percent to pretax profits.

The pie chart is also admirably suited for share-of-market presentations, with the forecasting company shown in the brightest color or shading to emphasize its position. Sometimes two pies are used—usually when the aim is to illustrate changes in share makeup over two periods of time.

4. The area chart.   This seldom used but often effective approach involves the use of geometric figures, drawings, or pictures—where the area of the illustration rather than its length or height signals its magnitude. The big plus here, of course, is dramatic impact. But unfortunately it is considerably more difficult to compare areas than heights—a physiological fact of life which can sometimes result in misleading interpretations (see below).

5. The statistical map.   This is a graphic device for depicting numerical information on a geographic basis. The way a map is shaded or hatched, for example, can be used to signal the magnitude of the phenomena under analysis or study. Rainfall maps are prime examples of this approach. Levels of wealth, population, unemployment, etc.—indeed anything that varies by geographic area—is amenable to this kind of illustration.

Dot maps are one variation. Here, the greater the number of dots in any area, the greater the concentration of the variable being measured. Another variation, the *pin map,* is often used to show locations of branch offices, plants, or retail outlets.

One drawback of the map approach: The area sizes depicted on the chart are often irrelevant. Thus, if retail sales are greater in New York than Arizona, this is not due to the fact that New York is larger in area, but rather to a population differential—something that is not readily apparent from the map itself.

6. The ratio chart.   One special variation of the line chart deserves further comment: the so-called ratio diagram. The chart, based on logarithms, is set up so that equal percentage changes show up as equal distances. This can often convey much more useful information than the more familiar arithmetic chart, where equal absolute differences show up as equal distances.

For example, if in one year sales went from $10 million to $13 million (a 30 percent gain) and in the next year they went to $16.9 million (again a 30 percent gain), the log chart would show an equal rise for both years. But if the conventional arithmetic chart had been used, the second year (with a $3.9-million rise) would have shown a sharper rise than the first year, for which only a $3-million boost was recorded.

Which is right? Both can be—depending on what one wants to measure.

The point to remember is this: When the purpose of a study is to analyze relative change rather than absolute change, the ratio chart is the proper one to use.

While the ratio chart makes use of logarithms, one doesn't have to be a mathematical genius—or, in fact, know very much about logarithmic computations—to use it; log graph paper, which does all the work for you, is readily available. All you need do is plot the data in the usual way on this log paper. The result is a ratio chart.

A simple example will illustrate its use. Suppose you are interested in plotting the sales growth of your firm for a five-year period:

Year	Sales
1966	10,000
1967	20,000
1968	40,000
1969	80,000
1970	120,000

The results on both log (ratio) paper and the more conventional arithmetic grid are shown in Figure 3-2.

FIGURE 3-2 *Log versus arithmetic scale.*

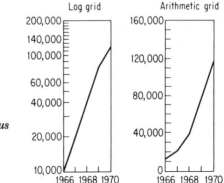

Notice that on log paper the line flattens out in the last year. That is because in the first four years sales were doubling (a 100 percent growth rate), while in the last year they showed only a 50 percent gain (a 40,000-unit increase over an 80,000-unit base).

Note that when arithmetic paper is used, the line over the last two years reveals no change in steepness. That is because there was an absolute 40,000-unit gain in both 1969 and 1970, which shows up the same on an arithmetic scale.

Other key areas where ratio charts are useful are:

(a) COMPARISONS: Many times, the basic problem is to compare trends where one variable is of small magnitude and the other is of large magnitude. To compare these on arithmetic paper gives a misleading picture. Only relative performances make sense—hence the use of the ratio chart.

(b) PREDICTIONS: Many business trends are complicated and hard to pinpoint. When they are plotted on log paper, they tend to become simpler (often turning into straight lines) and hence easier to work with and interpret.

(c) SPEEDY INFORMATION: Percentage changes can be obtained directly from the ratio chart without any arithmetic calculations, for it can be shown that the steepness of the plotted line at any point on a ratio chart represents nothing more than rate of change.

But for those who have an aversion to working with ratio charts, mathematicians have provided an out. Instead of plotting the absolute figures, the rates of change are plotted on an arithmetic grid—and the same basic effect is achieved. The calculations may be a bit more laborious, but, on the other hand, an arithmetic grid is less likely to frighten off some of the less sophisticated lay persons.

Distortions

Just as numbers can be misleading, so can visual representation of these numbers in the form of charts and graphs. The obvious example, of course, involves the direct plotting of distorted data onto a chart. Indeed, this compounds the felony because the visual approach tends to play up any misleading information.

But for those who want to distort, it is possible to do so even if the underlying figures are reliable and accurate. All that need be done is to present the data in such a way as to help the reader reach the desired (and often misleading) conclusion. It is the intention of this section to focus on this second type of misrepresentation—the kind that can be traced back to the choice of chart rather than to any inadequacies in the basic data. Some of the more notorious examples of graphic distortion are discussed below.

1. *The zero ploy.* In this situation the message can be altered significantly by changing the position of the zero line. Suppose, for example, that new orders which range, say, between $19 million and $20 million a month are plotted. If the new-order scale is started at zero, the changes look very small. That's because nineteen-twentieths of the available plotting space will be left blank—with all the existing fluctuations squeezed in the one-twentieth of the chart between $19 million and $20 million. Not a very exciting picture.

On the other hand, if the scale starts at $19 million and ends at $20 million, the resulting picture will be one of sweeping variation— swinging all the way from the bottom of the chart to the top. Dramatic? Yes. But it could also be misleading if the illustrator doesn't call attention to the fact that the picture presented is a magnification of only what is happening in the limited $19-million to $20-million range.

THE BIG QUESTION: How can you have your cake and eat it too? Again statisticians come up with a viable solution. In essence, it is a sort of compromise which reduces the possibility of misinterpretation, yet at the same time presents the message in eye-catching, dramatic terms. This is achieved by means of what is popularly known as a *scale break.*

This is usually accomplished by the use of the wavy line on the scale (or sometimes through the whole chart). Thus:

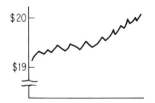 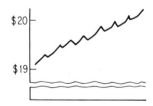

Unless attention were called to the scale break, the chart would give the impression to the uninitiated that orders fluctuated all the way from near zero to $20 million—an unwarranted conclusion in light of actual performance, which shows only about a 5 percent fluctuation.

Sometimes the break is omitted inadvertently rather than because of any premeditated desire to deceive. Indeed, such notably objective statistical agencies as the U.S. Commerce Department every once in awhile forget to insert a zero break in the charts they put out—though it is clear the agency has no particular ax to grind.

Unfortunately, this is not always the case. In some instances the zero break is omitted purposely—to guide the reader to a wrong conclusion. Politicians are particularly prone to this ploy. Thus a would-be candidate might plot the cost-of-living index (rising, say, from 100 to 103 in a given year) without using a zero break. His basic aim: to create an impression that inflation is rampant and that a change in officeholders is urgently needed to rectify the situation. A more objective presentation (with a zero break) would suggest that the index had gone up only 3 percent—a rather normal rise for this ever-climbing index of consumer living costs.

2. The area pitfall.   On occasion, charts showing area rather than height or length may be used to compare two variables or the same variable at different points in time. There is nothing intrinsically wrong or dishonest about this approach—except that as pointed out above, the eye is not usually trained to evaluate differences in area as accurately as differences in height.

Take the two circles below:

Obviously, the circle on the right is larger than the one on the left. But how much larger—10 percent, 50 percent, or 100 percent? Actually, it is 300 percent larger (four times as large) in terms of area. The tendency, however, is to underestimate the difference because the height of the circle on the right is only twice that of the one on the left.

It is even more difficult to compare accurately the volumes of two figures. Here, three dimensions (width, height, and depth) must be appraised—a difficult task even for the person with an experienced eye. Hence, they should generally be avoided when showing data graphically.

Unfortunately, the propensity toward underestimating area diagrams is sometimes exploited by unscrupulous statisticians. Thus a corporation anxious to play down a drop in earnings might designate earnings by a series of dollar bills of different sizes—the area of which represents profits. There's nothing illegal in this. But the planned effect—that of drawing attention away from the magnitude of the decline—has been achieved.

Sometimes the area approach is pursued because of its stronger pictorial impact. For example, it is quite effective visually to depict a doubling in housing by showing one house that is twice the area of the first one. On the other hand, using the "housing picture" approach without area diagrams would necessitate the use of two houses of the same size to denote a doubling in activity. This isn't nearly as impactful or pleasing to the eye.

Perhaps the solution is to label each area picture with its corresponding number equivalent. In this way impact is maintained, while the possibility of misinterpretation is reduced substantially.

3. *Absolute versus percent change.* As pointed out above, when rates are being measured, it is the ratio-ruled chart rather than the arithmetically ruled one that is required. The natural inclination to use the latter can sometimes lead the reader to a wrong or distorted conclusion.

Assume for a moment the plotting of GNP data over the past 150 years. The results on both arithmetic and ratio grids are shown in Figure 3-3.

Note that looking at the conventional arithmetic grid would lead one to believe that the GNP is increasing much faster today than it did, say, 50 or 100 years ago. The ratio chart, on the other hand, suggests a stable, or straight-line, increase over the entire period.

Which one is correct? Both are, depending on what story one is trying to get across. The arithmetic grid states that dollar growth today is bigger than it was, say, 100 years ago. This obviously has to be true since today the industrial base is so much larger than it was a century ago. The nation would be in pretty sorry shape if it couldn't rack up a bigger dollar GNP gain today than it did 100 years ago.

The second, or ratio-type, chart indicates that the rate of growth has been relatively stable over the period being illustrated. Thus if one is looking for overall performance over the longer pull—and this is generally the purpose of this kind of chart—then the ratio approach is the pertinent one.

The above suggests this rule of thumb: If only short-term dollar data are involved, an arithmetic grid is usually adequate since the month-to-month or year-to-year dollar changes are the significant points as far as judging performance is concerned. But when long-run data are

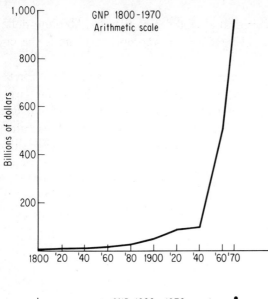

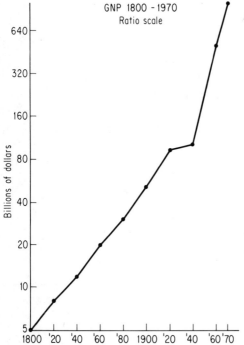

FIGURE 3-3   *Arithmetic versus ratio scale.*

involved, the emphasis is generally on the growth rate. Absolute dollar change plays a secondary role. Hence the appropriate chart is the one with a ratio-type grid.

As pointed out above, the ratio chart can also prevent misinterpretation when two series are being compared—one of which is far larger than the other. For example, plot the sales of a multimillion-dollar firm against those of a small one on an arithmetic grid and the fluctuations of the smaller corporation are lost (because the scale is too insensitive to dramatize each small dollar change). This might result in the erroneous conclusion that the large company's performance is more volatile.

Again the use of a ratio approach prevents this impression because both firms are compared on the basis of rate of change. If both move up or down by the same percentage, this shows up as an equal space shift on the chart. Put another way, the ratio grid facilitates comparisons of company performance when the companies are of uneven size. Nor need this be limited to comparison of companies. Comparisons of different products, different geographic areas, different salesmen's performance, etc., are also ideally suited for the ratio approach.

4. *The selective tack.* It is also possible to slant a message by illustrating only those points which are favorable to one's overall argument. In most such cases the reporter can plead innocence by pointing out that all the pertinent data are available in tabular form. What he neglects to say is that there was considerable bias in that part of the data which he selected to highlight via charts, graphs, or other pictorial material.

People in advertising and promotion often rely on this type of deception. A magazine that is increasing its advertising is quick to point this fact up in the form of a rising sales line in a chart. The fact that the magazine's competitors may be rising at a faster clip and that the publication is losing ground from a share-of-the-market point of view is conveniently forgotten when a decision is made on what points need illustration.

In a similar vein, the choice of color in a chart may be guided by the overall effect it is likely to have on the audience to whom it is directed. Specifically, bright colors may be used to point up favorable developments, while relatively neutral black and gray tones are used to illustrate less impressive aspects of a firm's operations. Along the same lines, charts revealing favorable results can be made more impactful or dramatic than those depicting less impressive information.

FOUR

Special Analytic Charts

The visual material discussed so far has for the most part been of the descriptive variety—giving a vivid graphic picture of what was happening, but in no way presenting specific action-oriented instructions for decision makers.

But much of this latter type of information and intelligence can also be presented in chart form. Indeed, in some cases, solutions to knotty business problems can often be read directly off a graph. Some of the most popular of these so-called analytic charts are discussed below.

Nomographs

These highly specialized charts are comprised of a series of scales placed side by side. Generally, if a line (or lines) is drawn through several of the scales, it is possible to obtain a computed result by extending the line (or lines) through still another scale.

Construction is based on the mathematical relationship between the pertinent scales—and generally (though not always) involves the use of logarithms. The big advantage is that these nomographs can be used by industry to get "quick and dirty" answers to many common families of problems.

There is nothing mysterious about how a nomograph works. Like any other chart, it is nothing more than a graphic representation of a mathematical relationship. Probably the simplest one to construct would be a nomograph for basic addition and subtraction.

This would be accomplished as follows: Take three arithmetic scales— equidistant from one other, but with the middle scale just double the value of the other two. Thus, opposite the number 4 on the two end scales would be the number 8 on the middle scale. Next draw a line from one value on the first scale to another value on the third scale. The reading where the line crosses the middle scale will be the sum of the original two numbers.

Since this kind of calculation is a simple task, no real savings is effected through the use of a nomograph. However, where complex multiplications and divisions are involved, nomographs with logarithmic scales can provide quick, "read off the chart" answers.

Figure 4-1, for example, is designed to provide clues as to which specific plastic material will yield the lowest overall cost in making a thermoformed skin package. The difficulty is that there are half a dozen frequently used plastics (polyethylene, polystyrene, polyvinyl chloride, cellulosics, etc.) that can technologically fill the bill. Note, too, that film thickness as well as film price affects cost.

Which to choose? The solution can be read directly off the sheet/film cost calculator shown in Figure 4-1 by simply following the directions given at the bottom of the chart. In the example, if (1) plasticized polyvinyl chloride was being considered at (2) a cost of 52 cents per pound with (3) a thickness of 5 mils, then by following the directions it becomes apparent that the film cost per 1,000 square inches will be approximately 12 cents.

As noted above, this is merely a graphic solution of an algebraic problem. In this example involving plastic costs, for instance, the algebraic formulation is also given (see Figure 4-1). Specifically, the cost of $D = 0.036\ A \times B \times E$. The precise relationship should always be given on the face of the nomograph to aid the analyst in his assessment of the results.

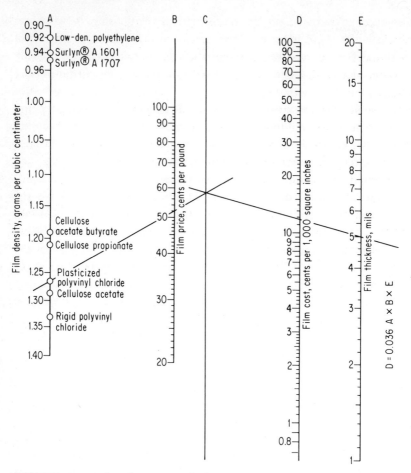

FIGURE 4-1 *Film/sheet cost calculator (Du Pont)*

Instructions: STEP 1: Locate the material on the density scale (scale A).   STEP 2: Locate the film price in cents per pound on the price scale (scale B). STEP 3: Draw a line between the points located in steps 1 and 2, and locate an anchor point as the intersection of this line with scale C.   STEP 4: Locate film thickness in mils of material being calculated on the thickness scale (scale E).   STEP 5: Draw a line between the anchor point on scale C (see step 3) and the point on the thickness scale located in step 4.   STEP 6: Read the film cost in cents per 1,000 square inches as the intersection of the line drawn in step 5 with the cost scale (scale D).

Scatter Diagrams

This is a technique used to ferret out the basic relationship—if, indeed, any exists—between two supposedly related sets of data. Assume, for example, a question involving the connection between imports and GNP. Theory would seem to suggest a positive relationship—specifically, that the more affluent a nation becomes, the more it is likely to import.

The first step would be to prepare a simple chart—plotting imports (vertical axis) against GNP (horizontal axis) as shown in Figure 4-2. Convention dictates that (1) the horizontal axis be used for the variable that causes the basic change (the independent variable) and (2) the vertical axis be used for the variable that is being influenced (the dependent variable). The final result is a series of points on the chart, each representing a different year. Since a large number of dots are scattered about the chart, this type of graphic presentation has come to be known as a *scatter diagram.*

A quick glance at the scatter diagram suggests that the data seem

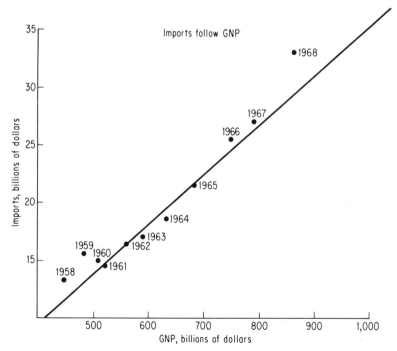

FIGURE 4-2   *Scatter diagram (U.S. Commerce Department).*

to follow a path. In short, there seems to be a definite relationship between the volume of imports and the GNP level.

The next step is to describe this path or pattern—in this case by a freehand straight line—since the dots do seem to approximate a straight line. Once this straight line is established, predicting becomes a relatively simple affair. Just take the GNP estimate (readily available from many sources, both private and public, for several years into the future) and read the estimated import level right from the chart.

Thus if the GNP were expected to hit, say, $1 trillion, we could predict with a fair amount of certainty that imports at that time would probably run something slightly above $35 billion.

In many cases where curved relationships are involved or where the analyst goes on to calculate the exact degree of relationship between the variables (the line or curve of regression), the freehand line is replaced by what is known as the *mathematically best* line. This is calculated by the method of least squares—a technique which ensures that the squares of the deviations about the line are at a minimum. This will be discussed more fully in Chapter 7.

The Break-even Chart

Many firms use a graphic approach to determine at what operational point they cease to make money and go into the red. In its simplest form this chart compares total revenue (based on the going market price) with total cost at differing levels of production.

The approach can best be understood by referring to Figure 4-3 and more specifically to the intersection of the total revenue (TR) and total cost (TC) curves. Note these key points:

1. Total revenue always rises by a constant amount—the price of each additional production unit.

2. The TC curve will start at a higher level than the TR curve (because there are fixed costs even at zero output).

3. The TC curve will end up at a lower level than the TR curve at reasonably full production. If it didn't, there could never be any profit in this particular production operation.

Figure 4-3 is based on an example where fixed costs are $100 million and variable costs are $5 million per 1,000 units turned out. Further, the price is set at $10,000 per unit. Note from the chart that the break-even point occurs at 20,000 units, or when revenue reaches $200 million.

Anything above 20,000 units means profit, and anything under 20,000 units means loss. Indeed, the exact profit or loss at a given production level can be read directly off the chart.

Thus the break-even chart enables a firm to set sales targets in line with overall profit goals. Also, by using different prices (and hence different TR lines), the firm can compare how well off it would be at varying prices that it might charge.

The technique is also helpful in determining the feasibility of future capital spending—for such investment can have both an upward and a downward effect on the break-even point. The upward push, of course, comes from the higher fixed costs that are likely to ensue. Assume for a moment that capital expansion superimposed on the example given above raises fixed costs by another $50 million. This in turn also raises the level of the TC line by $50 million, which then automatically advances the break-even point to 30,000 units. For only at 30,000 units will total revenue ($300 million) equal total cost ($150 million fixed and $150 million variable). It thus becomes clear that sales would have to increase another 10,000 units before the company would begin to show a profit.

But there is usually a concurrent downward pressing effect on the break-even point when capital spending goes up. This stems from the

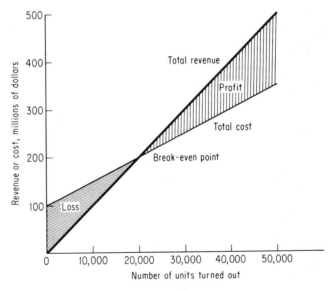

FIGURE 4-3 *The break-even chart.*

fact that the new plant and equipment are almost sure to introduce more advanced technology, which in turn will lower labor and other variable costs.

Referring to the above illustration again, if variable costs were to drop to $2½ million per 1,000 units (because of a $50-million increase in fixed cost investment), the break-even point would remain unchanged at 20,000 units. In this case the higher fixed cost would be just offset by lower variable costs. In short, it is necessary to assess all pertinent cost data before concluding any effect on the break-even point.

Note that as a rule of thumb, if the break-even point can be held, capital expansion is usually advisable, for the following reasons: (1) As sales rise above the break-even point, the profit potential under the postinvestment assumption becomes substantially more than would develop under the original situation, and (2) the greater capacity stemming from increased investment opens up the possibility of higher sales and hence higher profits.

The Learning Curve

This is essentially a visualization of the commonsense economic law that the unit cost of a new product decreases as more units of the product are made. Obviously, the price should be lower for each successive unit as the supplier, through the repetitive production process, learns how to make the product cheaper.

For example, the more times a worker repeats an operation, the more efficient he becomes in terms of both speed and proficiency. This, in turn, means progressively lower unit labor costs.

It also means lower costs in other ways. As an operation becomes familiar, there are usually fewer rejects and reworks, better scheduling, possible improvements in tooling, fewer engineering changes, and more efficient management control.

An illustration of a learning-curve application follows. Suppose that a producer finds he needs 100 units of labor to turn out the first item on a new product.

The producer further finds that the second item takes 80 hours to make—so the average labor cost for the two items is $\frac{180}{2} = 90$ hours. When production doubles to four units, the third unit requires only 73.9 hours to make; the fourth requires 70.1. So the average labor for all four units made to date drops down to 81 hours.

If 64 units were made, the learning trend would indicate that the vendor had learned to make them at an average of 53 hours for all units to date. If 128 units were produced, the average number of labor hours would be about 48, as shown by the falling curve in Figure 4-4.

Each time production doubled in the above example, the labor cost declined by 10 percent—thanks to the learning process. Thus the product is said to have a 90 percent learning rate, or a 90 percent learning

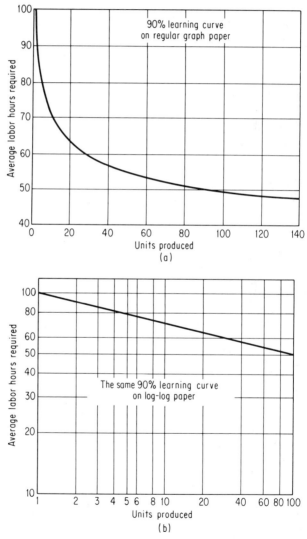

FIGURE 4-4 *Learning curves.*

curve. Put another way, this means that every time the total quantity of units doubles, the average time required to produce the new total quantity (old plus new) is 90 percent as great.

The basic thing to remember is that the learning-curve technique says that a given percentage reduction in costs results with each doubling of the number of units produced. This specific learning rate—a constant percentage per doubling of the number of units produced—occurs with regularity in many different industries.

Studies indicate that learning rates of from 75 to 95 percent are typical in the aircraft, electronic, and small electromechanical subassembly fields. However, learning curves can vary anywhere within the limits of 50 to 100 percent. The upper limit (100 percent), of course, implies no learning: the lower limit (50 percent) implies zero cost for the extra units—obviously a physical impossibility.

Note from Figure 4-4a how the curve flattens out as the number of units increases. This reflects the gradually diminishing effect of learning. After a few thousand units, the reduction in cost would become negligible. That's why learning-curve analysis is valuable only for new products.

When actually working with the curve, most analysts prefer to plot the data on special logarithmic scale paper (log-log), as shown in Figure 4-4b. There's nothing mystifying about this. All it does is convert the curve into a straight line to make it easier to work with. Indeed, cost reductions and estimated prices can be obtained by merely reading figures off this logarithmic chart.

To verify the fact that both charts (arithmetic and logarithmic) represent the same thing, look at the number of hours needed to produce 100 units on the upper and lower charts. Both indicate the same answer—about 50 hours.

The learning curve can also be used to solve other key business problems, including those discussed below.

Make or buy. A comparison of learning curves for your firm and your vendor can help you decide whether to make or buy, particularly when your plant is operating well under capacity. Your firm may have a faster rate of learning—but because your vendor is higher up on the learning curve, his costs will tend to fall faster than yours. This could possibly result in lower average costs for the number of units needed.

Estimating delivery times. Since the learning curve can forecast labor time required, it's possible to estimate how many units you can turn out over a specified time with a given labor force. For example, assume that you're making a complex electronic component and have the labor-force equivalent of 100,000 hours per month. If in April the average labor component (as determined by the learning curve) is 5,000 hours and in May it will drop to 2,500 hours, then you can turn out 20 components in April and 40 components in May.

Vendor progress payments. Since the learning curve suggests changing costs, it provides a basis for figuring a supplier's financial position on any given number of units. This is important because suppliers often operate in the red over the initial part of the production run, until learning can reduce costs below the average price. Buyers can minimize vendor hardship by using the learning curve to break down an order into two or more contracts—each with successively lower average prices, or they can set up progress payments based on the vendor's costs.

But like all statistical techniques, learning-curve analysis is not without its pitfalls. In particular, the following are things to look out for.

1. *Nonlinear relationships.* Learning-curve analysis requires a straight line on log-log paper. If the cost data don't approximate a straight line, the learning-curve approach described here won't work.

2. *Low-labor-content items.* Learning is most rapid when applied to labor—particularly assembly. On the other hand, if most work on a new item involves machine time, there isn't much opportunity for learning.

3. *Small payoffs.* Getting historic cost data for a learning curve involves much time and effort—particularly when a supplier uses a standard cost system. Therefore, learning-curve analysis is worthwhile only if the amount of money saved is substantial.

4. *Wrong learning rates.* Learning varies from industry to industry, from product to product, and from part to part. Applying one rate just because somebody else in the industry has used it can give misleading results.

5. *Established items.* If the vendor has made the item for someone else before, don't use the learning curve—even if the product is nonstandard and new for you. Most of the learning has already been done on previous work, and any additional cost reduction will probably be negligible.

6. *Misleading data.* Not all cost savings stem from learning. The economies of large-scale production may reduce costs, but this could hardly be described as learning.

7. *Nonproduction uses.* The learning curve is intended only for use in gauging costs of new production processes. Application to modification or other types of work could again produce misleading results.

8. *High learning rates.* Question any rates of 65 percent or under, for such rates are extremely fast.

Rate of Return on Investment Curves

When faced with a capital investment decision (whether to invest or not or which of alternative investments to choose) the decision depends in large part on the expected rate of return on investment (ROI). This is generally an extremely complex calculation—depending on many variables, including the expected income over the life of the new investment.

Shortcuts, however, have been worked up. One of the most popular is the series of curves derived by the Machinery and Allied Products Institute (MAPI). The MAPI approach makes some standard assumptions about future earnings of the asset or assets based on the past experience of metalworking companies. This, in turn, enables the analyst to apply some simple charts to come up with rate of return estimates. The charts are presented in Chapter 8, where there is also a more complete discussion of the investment decision.

Quality Control Curves

In judging whether a production run is up to snuff and whether it should be accepted or rejected, considerable use is made of two types of graphs: *control charts* and *operating characteristic* (OC) *charts*.

The former involve a graph containing two horizontal control lines. These indicate the range within which samples can be expected to fall on the basis of pure chance. Successive samples are then plotted as dots on the chart. Whenever a dot falls outside the control limits set by the two horizontal lines, it is said to be out of control—requiring a closer look into the factors behind the observed variation.

An OC chart is a bit more complex, but also involves the laws of probability. It consists essentially of a line or lines telling both producer and buyer the risks involved in accepting a given acceptance-rejection sampling plan. Examples of both the OC and the control charts—along with more detailed explanations—are given in Chapter 8.

Action-oriented Guidelines

The graphic approach can also be used to rate overall efficiency or performance. This is generally achieved by plotting actual operations or results against what might be expected on the basis of some time-tested yardstick or guidelines.

One company, for example, reports considerable success in plotting actual inventories against a so-called optimum stock level based on future production schedules, expected lead times, price trends, etc. Periodic referral to this chart furnishes an early-warning system—signaling when stocks may be growing either top-heavy or dangerously thin.

This type of performance yardstick is applicable to almost any facet of business activity—man-hours utilized, costs, profits, sales, etc. In all cases, however, there is one basic requirement—the ability to build up a reliable guideline, or norm, that is sensitive enough to catch imbalances before they get out of hand.

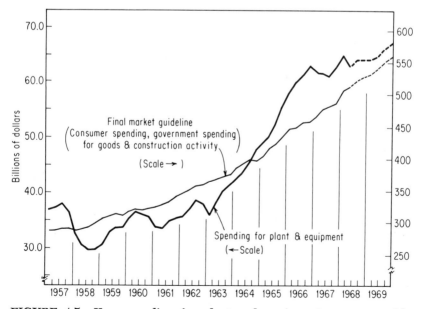

FIGURE 4-5 *How spending for plant and equipment compares with final market guideline (annual rates, seasonally adjusted) (U.S. Economics Corp.).*

Nor need this approach be limited to company performance. It is equally useful in developing guidelines which can provide early warning of changes in the business climate—changes which must eventually affect individual company as well as general business activity.

Thus the United States Economic Corporation, one of the nation's leading consulting firms, has developed an entire stable of these guideline indicators—covering such key economic areas as retail sales, inventories, and capital spending. Figure 4-5 illustrates how by following the capital spending guideline, the 1967 falloff in capital outlays would have been apparent more than a year before it actually got under way. Specifically, note how final market chew-up (a measure of how much capital spending is needed) started lagging well behind actual capital spending by early 1966.

The Useful Ogive

Many of the applications of cumulative-type curves were discussed in the previous chapter. Certainly, a chart pinpointing the number of people under twenty-one years of age or the number who are making over $15,000 a year can provide valuable information to marketing executives and government planners. But over and above these normal usages, the ogive has other special analytic attributes.

One of these involves the so-called Lorenz curve—and can best be described by way of a simple example. Assume that the problem at hand is to measure the equality or inequality of family income distribution in the United States.

The first step would be to divide the family units into subgroups in ascending order of income received. Thus the lowest 20 percent on the income ladder would be related to the percentage of the total United States income they received. The second 20 percent would be related to their income share, etc.

The next step involves cumulating up. The first point on Figure 4-6 would then involve the percentage of total income received by the lowest 20 percent on the income scale. The second point would cover the percentage of income received by the lowest 40 percent on the income ladder, etc. The last point, of course, would be the percentage of income received by 100 percent of the families—and obviously would turn out to be 100 percent of total income.

The points are joined to obtain a curve. Then by comparing this curve with a dashed diagonal line (also shown in Figure 4-6), one is able to ascertain the degree of income equality. The closer the curve is to the diagonal, the more equal the income distribution is. The extreme case (where the actual distribution coincides with the diagonal line) would imply that each segment of the population was receiving the same proportion of the total income. In other words, everyone would be making the same as everybody else.

The chart shown for the United States, incidentally, suggests that American income isn't nearly as unequal as some of this nation's detractors would have us believe. Indeed, if a similar curve were shown for most other industrial nations, the United States would rank very near the top as far as relative income equality was concerned.

Note, too, that the above illustration divides the population into five equal subgroups. But the same approach could have been used with a more minute type of breakdown—say, by deciles (10 percent subgroups) or even percentiles (1 percent subgroups). It all depends on the amount of detail needed.

Before leaving the subject of Lorenz curves, it should also be pointed out that they are often useful for comparing degrees of income concentration over different periods of time or under different conditions. On

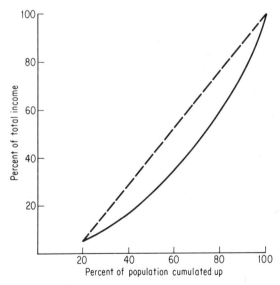

FIGURE 4-6 *The Lorenz curve (U.S. Bureau of the Census, 1965 data).*

the latter score, it is possible to draw two Lorenz curves on a chart—one representing before-tax income, and the other after-tax income. The extent to which the latter approaches the straight line of equal distribution measures the extent of progressivity inherent in the tax system.

Another type of ogive application involves its use in determining whether or not a particular set of observations can be described by the so-called normal distribution—a curve that is used extensively in sampling (see Chapter 5). Again a simple example can best illustrate the principle. Assume that sales of a product vary with income; the product appeals most to middle-income groups and has increasingly less appeal to people on both ends of the income ladder. The question is, Does a normal or chance curve best describe consumer behavior toward the product in question?

Plotting the raw data on an arithmetic grid can serve as a crude guide when marked skewness or asymmetry is present. But a much better test of normality involves (1) cumulating the observed data, (2) converting them into percentage form, and (3) plotting the resulting percentage series on ratio or semilog paper. If the resulting curve on this ratio paper approximates a straight line, then one can be fairly certain that the distribution under study is close to normal.

Finally, an ogive is sometimes used to determine graphically the median or middle value of a distribution. The process is relatively simple: (1) Plot an ogive of the data, (2) divide the number of observations in half and locate that point on the vertical axis where the number of observations is calibrated, (3) draw a perpendicular from that point on the vertical axis to the ogive, and (4) at the intersection of the perpendicular and the ogive drop another perpendicular down to the horizontal axis where the values are calibrated. The reading on the horizontal axis at that point is the median value.

FIVE

Sampling:
A Statistical Must

The technique of making an estimate of the characteristics of a large mass of data by examining a sample is an old one. It is only today's more scientific approach that is new.

The cook who sips the soup and the textile buyer who pokes his fingers into a bale of cotton are both putting their trust in the sampling method. They can be fairly certain of their conclusions, primarily because the products they are sampling are basically homogeneous; that is, they are subject to little, if any, variability. One spoonful of soup, for example, is pretty sure to taste like the next one.

The problem, however, is a little different when it comes to sampling business or economic data. If the responses of 10 prospective customers are examined, these can hardly be inferred to be exactly the same as the responses of the thousands of other customers who make up the overall market. The difficulty, of course, is that not every customer is alike.

More often than not, customers are likely to show wide variation, and by no stretch of the imagination can they be described as a homogeneous group. Ergo, a small, simple sample may well lead to erroneous inferences about the average customer and hence about total sales potential.

Similarly, it is hard to say with certainty that if one of ten sampled items proves defective, it follows that 10 percent of the entire production run is defective. Perhaps the sample just happened to pick the only defective item in the entire production run. One can never be sure.

Why sample, then? Why not play it safe and collect all the pertinent data? Or, as statisticians put it, why not study the entire population or universe? There are several good reasons, discussed below, for avoiding full enumeration of data—some of them are quite obvious, and others not so obvious.

1. *Cost.* Clearly, if a sampling of 5 or 10 percent of the data can give a reliable estimate of the population or universe, a tremendous savings can be effected. Often this is a must because appropriations just won't permit the collection and tabulation of all the available statistics. It's either sample or forget about the entire project.

2. *Practicality.* In many cases it is virtually impossible to collect all the data, even if the money should be available. For example, counting the total number of unemployed in a given month just can't be done. For one thing, it would be too large a job to complete in the relatively short period of a few weeks. Second, no matter how careful one was, a goodly number of the unemployed would escape the statistical dragnet. They just wouldn't stand still long enough to be counted.

3. *Waste.* It just doesn't make sense to use data that are more precise than needed for the immediate purpose—particularly when the increases in precision do not make the data any more useful. Thus it would make little difference to a seller of a particular brand of soap whether 84, 85, or 86 percent of the customers liked its smell. If the percentage were that high, the seller would in all likelihood continue to market this product.

4. *Destruction.* In many instances a sample is necessary because the product, in the process of being measured, is destroyed. Perhaps the best example here would be that of a manufacturer who wants to test light bulbs to see how long they last. If he tests all of them, he'll have none left to sell. It's either sample or forget about getting an estimate of the bulbs' life.

5. *Spurious accuracy.* Try counting the number of marbles in a large glass urn several times. More likely than not, each count will yield a different answer because of the pure drudgery and boredom involved in such an enumeration. An equally accurate estimate could probably be obtained by counting the number of marbles in 1 cubic inch of the urn (a sample) and then multiplying the result by the total number of cubic inches the urn contains.

In short, it is clear that sampling offers tremendous advantages—and in many cases is the only feasible approach to measuring some aspects of business data. Once recognized, the problem boils down to selecting a representative group or sample out of a large population or universe. This isn't always easy. Nevertheless, it is possible—provided proper precautions are taken.

To be sure, such a sample won't yield precisely the same results as a full count of the entire population. But if drawn in a scientific manner, it is possible to compute the probable range of sampling error that might be involved. And if the error is small enough, the sample can readily be used as an approximation of the entire population.

In any case, it should be clear that a businessman, more often than not, has to make decisions based on something less than complete data. Basically, his problem is (1) to assess the limited evidence available, as to both accuracy and reliability, and then (2) to make the decision that has the best chance of optimizing his particular goals.

Of course this sampling, or probability, approach can lead to a wrong decision. There's no 100 percent guarantee. Rather, the aim of sampling is to (1) accumulate representative information in a hurry, (2) assess the chance of error, and then (3) steer the decision maker onto the course of action with the most likely chance of success.

Sample testing of, say, 1 percent of the widgets coming off a production line might let us know with about 99 percent accuracy whether to accept or reject the lot. The businessman might well be willing to accept the 1 percent risk of making the wrong decision in return for the tremendous potential savings derived from being able to differentiate most of the time between good and bad lots. It's a risk, yes. But it's a calculated risk—one that is likely to pay off over the longer pull, with the number of times the analyst guesses right far outweighing the one or two times he makes the wrong decision.

Sampling the Data

The first basic problem in sampling involves basic common sense rather than any advanced degree in mathematical statistical theory. Specifically, How does one go about gathering a sample? Obviously, the sampling units—whether they are customers, widgets, or prices—must be representative of the population or universe to be measured.

To obtain such a representative sample is easier said than done. Indeed, it is often quite difficult to avoid some kind of bias—in either the collection or the interpretation phase or the sampling procedure.

The classic example of how a sample can go awry is the one involving a Presidential preference poll conducted more than 30 years ago. The time was 1936, and the contestants were the incumbent Democrat (President Franklin D. Roosevelt) and the challenging Republican (Alfred M. Landon, of Kansas).

The *Literary Digest,* the magazine taking the poll, sampled a supposedly statistically significant number of voters by telephone. Their prediction: Landon by a landslide. History, of course, proved the *Literary Digest* woefully wrong. It was Roosevelt who won by a landslide.

Where did the *Literary Digest* go wrong? It was a simple case of not obtaining a representative sample. Remember that the voters polled were telephone owners—a group which at that time were a wealthy (and hence scarcely representative) portion of the population. Inferring total population behavior on the basis of a few wealthy individuals was clearly unjustified. Incidentally, the *Literary Digest* paid dearly for its error—folding up soon after the poll-taking fiasco.

Lack of representativeness can stem from less obvious mistakes, too. Take the simple problem of polling an industry to ascertain sales potential. The easy way out, of course, is to hit the few large firms. But this can lead to costly error. One equipment maker tells the story of having done precisely this and arriving at an estimate calling for a 15 percent spurt in his firm's sales over the coming year. Their actual results, however, showed a much more modest 5 percent advance.

Again, the trouble was traced to an unwarranted assumption—specifically, that because the large companies were planning on a 15 percent jump in their machinery purchases, the small firms would behave in the same way. As it turned out, the small companies at that time were caught in a profit squeeze—and were in no position to raise their capital spending sights by such a large amount.

Sometimes the problem of unrepresentativeness is subtle enough to escape attention. Consider the case of a firm that wants an estimate of the potential number of retail customers entering a store. If customers are surveyed only in the morning, this could well lead to an underestimate—since traffic reaches its peak in the afternoon. Again the moral is the same: A sample must provide a miniature snapshot of the parent population. If it doesn't, then all bets are off. How, then, can one assure that a sample will be representative? The first step is to design an approach that will (1) assure a random selection of items from the population and (2) reflect, when and if necessary, the heterogeneity of the data. One of the following procedures can usually prove adequate.

1. *The simple random sample.* This is generally used when the population is relatively homogeneous—say, the width of widgets coming off a production line. Under such conditions all that's needed is to make sure that each item in the population, e.g., each widget, has an equal chance of being drawn in the sample.

2. *The systematic sample.* This is a variation of the simple random sample. To speed up the selection process, the sampler chooses every nth item—say, every tenth in an alphabetical listing. There is, however, one danger in using this approach. It is important to avoid instances where that sampling interval (each nth item) coincides with a constantly recurring characteristic of the list. For example, if every tenth widget always happens to come from the same machine, then obviously the sample will not be representative of widgets coming from all machines.

3. *The stratified sample.* This is essentially a hybrid-type random technique used with heterogeneous data. The approach is to divide the population into strata or levels—each of which is more or less homogeneous—and then take random samples from each stratum. If, for example, the problem is to make a study on department store sales, the sample might be divided into strata consisting of stores grossing over $10 million, stores grossing $5 million to $10 million, etc. A random sample would be drawn from each stratum and then combined with proper weighting into an estimate for the population. It follows, then, that with more accurate estimates for each stratum, the estimate of the entire population should be more accurate for a given sample size.

4. *The cluster sample.* This is also used when there is a lack of homogeneity in the population to be sampled. But in this case the strata are unknown or hard to isolate. What is done, then, is to sample random areas or "clusters" of the population—and then treat each area or cluster

as a sampling unit. Thus if one wanted to make a survey of family incomes within a city, one might divide the city into small individual areas. Second, a random or systematic sample of clusters would be drawn. Then the clusters thus chosen would be sampled for family income, keeping in mind that the chosen areas or clusters rather than the families form the basic sampling unit.

One drawback: It can be shown statistically that for a given sample size, the cluster approach tends to yield somewhat more statistical variation than a pure random sample. But the larger sample size needed to achieve equal reliability is often more than compensated for by the fact that the cluster sample is generally a lot cheaper—because the sampling is done within tight clusters rather than being scattered over the entire universe or population.

5. Sequential sampling. This is an approach using relatively small samples. The technique is to start off with one small sample; if no decisive results are obtained, a further small sampling is taken. It can be shown statistically that sequential sampling involving a given number of observations yields more reliable results than one large sample of the same size.

Sometimes when the sample is derived from, say, a mail survey, a representative cross section of the population may be queried—and yet the results do not represent the population. That's because the type of people who respond may differ significantly from those who chose to ignore the survey. Thus the proportion of responses by ethnic group, age, income, etc., may not be in line with the proportions of these factors in the overall population.

As with the other such nonrepresentative results, the possibility of bias can't be ignored. The problem of nonresponse should be approached by spending more time and effort to reduce respondent resistance to a minimum.

But bias isn't always unintentional. Every once in awhile a survey will be designed to prove a point rather than to seek out whether a point is true or not. Happily, this kind of intellectual dishonesty does not occur very often. Nevertheless, it's something to look out for.

An example of this kind of bias might occur if, say, a cigarette company were trying to prove a college student preference for its own particular brand—brand A. The firm might distribute free samples of brand A, quickly following this with a campus survey on what brand the students were smoking. It would indeed be surprising if the students didn't "prefer" brand A at that particular point in time.

Size is still another problem in any sample-taking decision. Obviously, a large sample is going to be more accurate than a small one. Indeed, if one takes a large-enough sample, it approaches the population and hence 100 percent accuracy. The problem here, of course, is how much accuracy one needs and what price one is willing to pay for it. The statistical considerations involved in making this decision will be discussed below on page 97.

Another tough problem area involves the actual designing of the questions in a survey or sample. The way queries are posed, tabulated, and interpreted, for example, can often mean the difference between a go and a no-go decision on a particular product. Headaches here crop up on several fronts.

1. Respondent interpretation.   In general, surveys should be constructed so that the respondent has little doubt as to what the questioner really means. Thus, to assure consistent results when asking for age, for example, it is usually a good idea to qualify the query with, say, a request that the respondent round out his answer to the nearest year. In short, don't be ambiguous.

One common type of misinterpretation involves the use of "yes" or "no" choices—for example, a question such as "Do you smoke?" This type of question could give trouble to a respondent who would be inclined to answer "Sometimes, depending on the situation." A better way to phrase the question might be, "Do you smoke (a) always, (b) most of the time, (c) occasionally, or (d) never?"

2. Respondent distortion.   People don't always tell the truth—most of the time because they just don't know what the truth is. Thus if you ask a respondent whether he would like, say, a cake mix in a red and yellow package, he might well say "yes" because of the power of suggestion. A better way of obtaining the answer might be to ask, "What color would you associate with a cake mix?"

3. Classification.   Responses must be classified and tabulated in a meaningful manner. This would usually eliminate the "Why do you smoke brand A" type of question. A much better way to get at the underlying reasons is to submit a list of possible factors and then have the respondent check off the appropriate ones.

4. Statistical summation.   Here the problem is to take all the properly classified responses and convert them into some action-oriented yardstick. Thus the Michigan Survey Research Center prepares an index of consumer confidence every three months on the basis of a series of questions. Without its index the reader would be hard-put to evaluate the overall responses in one period as contrasted with those in another period.

5. Analysis. Even if a meaningful summary can be calculated, this still leaves the problem of "What does it all mean?" Again take the consumer confidence survey as an example. Does a 5 percent rise in consumer confidence suggest a 5 percent rise in retail sales? It might, and it might not. To get the answer to this, a study of retail sales and the consumer confidence index over a period of years would be necessary. And since other factors over and above confidence influence sales, these too would have to enter into any overall evaluation.

To sum up, taking a sample involves a lot more than just taking a certain percentage of the population and asking a few simple questions. Representativeness, bias, statistical adequacy, survey design, and interpretation are all serious problems that have to be tackled before any sample results can be used for dollars-and-cents decision making.

Sampling Error

Choosing the proper sample is only the first step—albeit an important one—in estimating the attributes of a given population or universe. That's because no matter how much time and effort have been devoted to the selection of a representative sample, there is still the possibility that the sample value will vary from the true population value on the basis of pure chance.

In other words, there are actually two types of error possible in sampling—one due to a slipshod choice of sampling units which results in a nonrepresentative sample and the other due to the inherent nature of the sampling process itself. This second kind of error or variation is unavoidable. Nevertheless, it must be recognized and evaluated if sampling procedures are to have any validity.

The basic difference between the first type of procedural error discussed in the previous section and the statistical sampling type alluded to here can best be illustrated by a simple example. Assume samples of two items from a hypothetical truckload of widgets, where one out of every two widgets is defective.

A procedural type of error might involve the selection of sample widgets from only a few rather than all of the machines which produced the widgets. If this is the case, it is not unlikely that the two-unit samples will yield something more or less than 50 percent defective.

A chance or sampling error, on the other hand, would occur even if the sample were 100 percent random—with each widget in the truckload having as much chance of being chosen as any other one. Assuming such a random selection, clearly the odds favor picking one good widget and one defective one when taking a sample of two—since each kind has an equal chance of being selected.

Nevertheless, there is a fair chance or probability that a variation from the population percentage may crop up. The reader may want to verify this by taking two-unit samples from a bowl containing, say, an equal number of black and white balls. He will quickly find out that there is a fair chance of picking two of a kind quite often.

The only meaningful way to deal with these three possible alternatives is to evaluate the chance of getting each of the three (two good ones, two bad ones, and the one-good–one-bad combination). It can be shown through some simple mathematics that the one-good–one-defective combination can be anticipated on the average about 50 percent of the time—with the chance of getting two good ones (or two defective ones) occurring about 25 percent of the time.

While the aim of this book is to eschew the mathematical approach, it may be useful in this particular instance to go into the actual calculations of the above percentages—for all evaluation of sampling error or chance variation (even in the most complex cases) is built upon the same basic probability principle.

The solution proceeds along these lines:

1. The probability of selecting a good widget when drawing the first item is one-half, or 50 percent.

The probability of getting two good widgets in the two-unit sample involves simply the multiplication of the probability involved in the picking of each unit. Thus $\frac{1}{2} \times \frac{1}{2}$ equals $\frac{1}{4}$, or 25 percent. (Obviously, adding instead of multiplying the probabilities would lead to an unrealistic answer because the probability would then be 1, or 100 percent, a situation which would immediately rule out the very distinct possibility of obtaining either of the other two alternatives—two defective units or a combination of one good and one defective unit.)

2. Similar reasoning yields a 25 percent probability of obtaining two

defective units when drawing a two-unit sample from a population lot containing 50 percent defective.

3. Subtract the probability of getting two good widgets (25 percent) and the probability of getting two defective units (another 25 percent) from 100 percent, and the remainder yields the 50 percent probability of getting the third alternative—the one-good-one-defective combination.

Unfortunately, if one is dealing with a sample consisting of hundreds of units instead of the two described above, this kind of approach can become extremely tedious and time-consuming. So again statisticians have come to the rescue with a mathematical formula that will yield precisely the same results. It is known as the *binomial expansion* or *distribution,* and it provides the pertinent probabilities via the expansion of the expression

$$(a + b)^n$$

where $a =$ probability of getting a good widget
$b =$ probability of getting a defective widget
$n =$ sample size

In the above example this would mean expansion of the expression

$$(\tfrac{1}{2} \text{ good} + \tfrac{1}{2} \text{ defective})^2$$

Expand (multiply the portion in the parentheses by itself), and the results are

$$\tfrac{1}{4}(2 \text{ good}) + \tfrac{1}{2}(1 \text{ good, 1 defective}) + \tfrac{1}{4}(2 \text{ defective}) = 1$$

In other words, there is a 25 percent chance of getting two good widgets, a 50 percent chance of getting one good and one defective widget, and a 25 percent chance of getting two defective widgets—with all the probabilities adding up to 100 percent, or unity. It's interesting to note that in this hypothetical example, only 50 percent of the samples can be expected to yield precisely the same percentage as the population. Clearly, then, chance variation can't be ignored.

It is from this sample binomial formula that all probability theory is built up. When dealing with large numbers of items in a sample (over 30), a very useful approximation of this binominal distribution—the *normal curve*—is a lot more practical. This bell-shaped curve (see Figure 5-1), like the binomial distribution, describes how samples can be expected to vary about a population measure on the basis of pure chance. Thus in the above widget sampling problem the normal curve, or *normal distribution,* as it is sometimes called, would tell us the probability of getting any specific sample percentage.

The downward-sloping trend on both sides of the population percentage (located at the apex of the curve) should also be noted. This

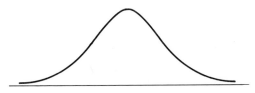

FIGURE 5-1   *The normal curve.*

suggests that the bigger the sample variation from the population reading, the smaller the chance of its occurring—with only a few samples straying significantly from the population measure.

The word "significant" is an important one in statistics, for there has to be a cutoff point for rejecting or accepting a sample. Again, a simple illustration can best point up the principle. This time assume that the long-run average population of defectives is 10 percent. Assume further that a sample yields a defective rate of 20 percent. The key question: Should the batch from which the sample was drawn be rejected as being of below-average quality?

The answer involves the determination of just what percent of times a sample of 20 percent defective can be drawn on the basis of chance when the true population percent defective is only 10 percent.

The precise solution depends to a large extent on the size of the sample and the inherent variation of the widget population. But once these variables are known, it is a fairly easy task to figure out just what the chances are of getting a sample of 20 percent defective.

If such chances are found to be very slight—say, one or five times out of a hundred—the conclusion would be that it is highly unlikely that the batch sampled has only 10 percent defective. The decision then would be to reject the batch of widgets from which the 20 percent defective sample was drawn.

In statistical parlance this approach is referred to as the *null hypothesis.* The logic is as follows:

1. Assume that there is no difference between the sample percentage

and the population percentage other than that due to pure chance factors.

2. Find out what the odds of getting such a difference are if only chance factors operate.

3. If such odds are sufficiently small, reject the assumption that the difference is due to chance—or, to put it another way, reject the null or no-difference hypothesis.

4. Rejecting the null hypothesis leads to the inference that there is a nonchance factor operating—and it is this nonchance factor that is causing the variation between the sample and the population percentages. Or—again using statistical terminology—there is a significant (nonchance) difference.

Generally, but not always, 1 percent and 5 percent are used as cutoff points to indicate whether variation between the sample and the population is due to chance or is significantly different. Some people use this rule of thumb: (a) If the probability of difference due to chance is more than 5 percent, then a significant difference is not proved; (b) if the probability lies in the 1 to 5 percent range, then the difference may or may not be significant; (c) if the probability is under 1 percent, the difference is clearly significant.

The indispensable vehicle for carrying out this approach is the normal curve. Tables have been set up allowing one to calculate relatively quickly and easily the precise chances of getting a specific sample deviation from the given population when only chance factors operate.

A note for the mathematically oriented: The sample-population differences presented in these tables are expressed in terms of commonly accepted units of variation known as *standard errors*. By converting the differences into the number of standard errors away from a population, the tables become independent of absolute units of measurement and hence can be used to solve any number of similar problems.

Before leaving this section, a word of caution may be needed concerning the word "significant." Thus far we have been talking about significant variation in the statistical sense—that is, when the variation is well beyond the likelihood of chance. Thus if a sample value can vary from a population value because of chance only 1 percent of the time, we would say that it is well beyond the likelihood of chance—or that the sample value is significantly different from the population value.

But there is another type of significance—practical significance—that must also play a role in any decision to accept or reject a sample. More specifically, a difference may be statistically significant but still be of no practical significance or importance. In this case to reject the sample would be foolish.

Again an example can best illustrate the principle involved. Assume the testing of cylinder width in a given production run, where it is known that specifications call for a 10-inch width. We may get a sample value of 10.09 inches, which is found to be statistically significantly different from 10. On the other hand, as a matter of practicality, the cylinders can be used as long as they are within 0.10 inches of the specification. Here we have a case of statistical significance but no practical significance.

The obvious decision here is to accept the production run even though it is significantly different from the population. In other words, there is probably a real difference, but in this case it is simply not worth acting upon.

Some Typical Applications

The evaluation of samples through the use of probability techniques has blossomed into a full-fledged statistical discipline over the past two decades. Indeed, the approach has been so expanded and refined that today there is scarcely a sampling problem that is not amenable to this rigorous type of evaluation and analysis.

The problem, however, is that much of the theory is based on advanced mathematical techniques. But for the busy business executive this need not prove insurmountable. For the need here is not so much for detailed methodology, but rather for knowledge of the types of problems which can be solved and the basic rationale behind the solutions. In other words, what this section aims to provide is the answers to two questions: (1) Where can I apply sampling theory? and (2) What do the results really mean?

The following list is not meant to be exhaustive. Rather, the aim is to present some of the more typical problems—the ones that are most likely to crop up in a business situation. Note, too, that they are all based on the laws of chance—and hence assume that the sample results are both random and representative.

1. *Sample versus population.* This is the broad type of problem described above. There are innumerable occasions—the problem concerning defective and nondefective widgets for example—when a deci-

sion has to be made on whether or not a particular sample comes from a specified population.

Aside from the area of quality, this sort of question might well crop up in market research. Thus a department store catering to a given income group may be thinking of opening up a branch in a new suburb. No census figures are yet available—but the store thinks the incomes in the suburb match those in its general customer profile. The store might then sample the income of families in the suburb—and make then a decision on whether any variation from its established market is significant or just due to chance.

Note, too, that the significance level may also be varied, depending upon the type of problem. Thus a computer maker would want 99.9 percent–plus assurance that a given production run is up to standard. On the other hand, a light-bulb manufacturer might settle for 95 percent certainty.

Another complicating note: In all types of sampling problems involving the null hypothesis, it should be kept in mind that two kinds of probability error are possible. The first involves rejecting a sample when in fact it should have been accepted. Say the chance of obtaining a difference as great as you obtained is only 1 percent. It is then possible (though certainly not very likely) that you may just have the odd-ball sample. If so, you would be making an error in rejecting it. Statisticians call this a *type I error*.

There is also the very real possibility of accepting a sample when in fact it should have been rejected. Thus our sample may come from a completely foreign population—but by pure chance it happens to fall close to the population we are concerned with. To accept it is to make a different type of error—one that statisticians call a *type II error*.

All statistical sampling must keep both types of error in mind. Usually cutting down on the chances of making one type of error tends to increase the chances of making the other kind. Which choice one makes depends on the type of error that is likely to prove most costly. Thus a parachute maker would be more concerned with a type II error—accepting a sample when in fact it should be rejected. In short, he just can't afford to run the risk of accepting a defective product.

2. Choosing between samples.   Many types of business problems are centered around making a choice between several different alternatives. Again probability theory can be called up to yield a meaningful answer.

Take the situation where two alternative processing techniques are available to a metal producer—both of which are supposed to increase tensile strength. The producer, faced with making a decision, might test samples from both processes. He would then (*a*) calculate the observed difference between both processes and (*b*) ascertain whether the difference was significant or simply due to chance.

The brand A–brand B preference study—used so glibly in advertising—provides still another example of where this approach may be used. Only this time, rather than using an experiment, the survey route to sampling is taken. Interpretation from the statistical angle, however, would be exactly the same.

Some organizations that do extensive sampling have set up handy reference tables which enable the analyst to know at a glance whether the two samples are significantly different. Table 5-1 is one such table (directions for its use appear below it).

For example, in samples of 2,000 where the average percentages are somewhere near 50 percent, the table tells us that any difference of more than 4 percent is statistically significant.

3. *Three or more samples.* Here a variation of the normal curve—called the *chi-square distribution*—can be of tremendous help. If, for example, there are three machines, each of which (upon sampling) yields a different percent defective, a question arises as to whether there are any significant differences between the three machines.

The easiest way to solve this problem is to table the actual number of defective and nondefective from each machine. Then this number is compared with the theoretical or hypothetical number of defective and nondefective parts which might have been produced if no real differences had existed between the three machines.

The variations between theoretical and actual magnitudes are then evaluated via the chi-square test, which provides the probability of getting the observed amount of variation if only chance factors operated. If the probability is low enough (say, 1 percent), then a decision might be made that there is a significant difference between the machines.

This particular test is known as a *3 × 2 contingency table* since three machines are being measured for their ability to turn out either defective or nondefective parts. But the problem need not be limited to three machines. If four are under study, then a 4 × 2 contingency table would be in order.

Nor need the problem be limited to an either-or classification (defec-

TABLE 5-1 *Determining the Significance of Sample Differences*

Size of samples compared	Approximate sampling tolerances for differences between two survey percentages at or near these levels, %				
	10% or 90%	20% or 80%	30% or 70%	40% or 60%	50%
2,000 and 2,000	2	3	4	4	4
1,500	2	3	4	4	4
1,000	3	4	4	5	5
750	3	4	5	5	5
500	4	5	6	6	6
250	5	7	8	8	8
100	7	10	11	12	13
1,500 and 1,500	3	4	4	4	4
1,000	3	4	5	5	5
750	3	4	5	5	5
500	4	5	6	6	6
250	5	7	8	8	8
100	8	10	12	12	13
1,000 and 1,000	4	4	5	5	6
750	4	5	5	6	6
500	4	5	6	7	7
250	5	7	8	8	9
100	8	10	12	13	13
750 and 750	4	5	6	6	6
500	4	6	6	7	7
250	5	7	8	9	9
100	8	10	12	13	13
500 and 500	5	6	7	8	8
250	5	8	8	9	9
100	8	11	13	14	14
250 and 250	7	8	10	11	11
100	9	12	13	14	14
100 and 100	10	14	16	17	17

(95 in 100 Confidence Level)

SAMPLING TOLERANCES WHEN COMPARING TWO SAMPLES

Tolerances are involved in the comparison of results from different subgroups of a sample and in the comparison of results between two different samples. A difference, in other words, must be of at least a certain size to be considered statistically significant. Table 5-1 is a guide to the sampling tolerances applicable to such comparisons.

When to use this table
Use this table when you need an approximate guide to statistical significance of a difference between two survey percentages. The question you usually wish to answer is, Is the difference in percentages great enough to place some confidence in the result?

The term "statistically significant" is used to refer to a difference larger than that shown in the table. That is, the researcher can be reasonably confident (at least 95 times out of 100) that it is a true difference and not due to chance alone.

Avoid using the table as a statistical "crutch"
The subject matter importance of differences cannot be measured by this type of statistical test. The statistical test is always auxiliary to the main questions: Is the result important? and What does it mean?

Very small and perhaps unimportant differences will be statistically significant if the samples are very large.

Differences which are perhaps very important because of their nature and magnitude may not be statistically significant if the samples are small.

SOURCE: ORC Caravan Surveys, Inc.

---

tive versus nondefective). One firm tells of suspecting that the frequency of absences was related in individual departments to the day of the week. Six departments were studied, with absences tallied in each department for each day of the week. Thus, a 6 × 5 contingency table was used.

4. Estimating the population.   There are still other problems where the need is essentially to obtain an estimate of the population or universe. The procedure here is to use the sample as the basis for making such an estimate.

A typical example that recently occurred in the advertising profession might best illustrate the approach. A regional firm was considering spot announcements on ultrahigh-frequency television—but before committing money, the firm wanted to make sure it was reaching a minimum of about 10 percent of the total families. So the firm decided to sample TV sets in its marketing area and found out that about 11 percent of the sets sampled were of the UHF variety.

On the basis of probability theory the company was able to say with 99 percent certainty or confidence that the true proportion of sets with UHF varied somewhere in the 10 to 12 percent range. This provided enough assurance for the manufacturer that it would achieve its minimum reach. So it decided to go ahead with its advertising plans.

Statisticians lump the whole family of problems involving this type of approach under the title of *confidence intervals*—since it involves making an estimate of the population with a given degree of confidence.

Many organizations that set up handy reference tables to ascertain whether differences between two sample percentages are significant (such as the one presented in Table 5-1) also work up tables which will permit analysts to estimate from a sample the probable range wherein the population percentage lies.

5. Correlation sampling.   A sample of consumers may reveal that a good relationship exists between spending and income. But is it due to chance? Again appropriate probability tables are available to provide the answer.

Assume that the sample correlation relationship is good. Then the next step would be to use a statistical test to see whether the good relationship is significantly different from a zero relationship, or nonrelationship. The null hypothesis is involved here too—with the assumption made that the sample correlation does not vary significantly from a zero population correlation. A test is also available to see whether one sample correlation is significantly different from another one.

Nor does the process stop here. The significance of partial correlation can also be measured statistically. Thus if spending were related to both income and assets, it would be possible to measure whether the income aspect was significant, the asset aspect was significant, or the combined influence was significant. This has become especially important in recent years with the advent of the computer, which makes it possible to experiment with 50 or 100 influences on spending—only a few of which may actually be significant. This process of choosing only the significant influences has been incorporated into a new technique known as *stepwise regression* (see Chapter 7).

6. Quality control.   The rejection of a particular sample described in the widget problem above only scratches that surface of the statistical quality control approach. Generally speaking, most of the techniques boil down to setting up a system of acceptance sampling. This involves inspecting a stated number of items from a shipment or production run. If the defectives are not more than a stated number, then the shipment is accepted. If, however, defectives exceed the stated number, two alternatives are open: (*a*) full inspection (called *acceptance-rectification sampling*), in which case all items are inspected and the defective ones are sent back to the supplier, and (*b*) lot rejection (called *acceptance-rejection sampling*), where the whole shipment is sent back to the supplier without any further testing.

Acceptance sampling can also be broken down into single, double, or sequential sampling. In the single variety, the first sample result provides the basis for decision. In double sampling a second sample is taken if the first one isn't conclusive as to whether acceptance or rejection is called for. Sequential sampling involves taking as many samples as needed to reach a decision. It's a takeoff on the basic sequential approach described above.

Determining the Sample Size

So far the discussion has stressed the physical problems of drawing a representative sample and the subsequent statistical procedures needed to process and evaluate the results. But one other key question remains to be answered before any meaningful procedure can be set up for practical business use—namely, How large a sample will be required?

Obviously, greater precision and accuracy can be attained by increasing the absolute size of the sample. Thus, statistically it can be shown that accuracy tends to increase with the square root of the sample size. If, for example, you want to double accuracy, the sample would have to be increased fourfold. If you want to triple accuracy, a ninefold increase would be necessary, etc.

The first problem, then, is to determine the degree of accuracy needed—and from that deduce the sample size necessary to attain this degree of required precision. The decision on the exact amount of precision needed is a subjective one in that some manufacturers might be able to settle for less or more accuracy than others. A project with more money at stake, for example, would normally call for greater precision. But in all cases the accuracy should be such that it permits action-oriented decisions to be made on the basis of the sample findings.

There are other considerations, too, which can affect sample-size requirements. It can be shown, for example, that accuracy will vary directly with the variability of the population being sampled. Other things being equal, a population with little spread or dispersion will tend to be more accurate and thus require a smaller sample size than a population with wide variation. For example, if one were measuring income in an all-wealthy neighborhood, a smaller sample would be needed than if one were measuring income in a neighborhood in which low-, middle-, and upper-income classes were represented.

Also, to some modest extent, the size of a sample, and hence its accuracy, may depend on the population value. Thus, a production run where normally about 30 percent of the widgets are defective would need a somewhat larger sample than a population where normally only 10 percent are defective.

On the other hand, it is interesting to note that sample size is only little affected by size of the population. Thus a sample of 500 from, say, a population of 50,000 is likely to prove little more accurate than a sample of 500 from a population of 500,000.

In any case, it is foolhardy to start a sampling program without being reasonably sure of the required accuracy. Too small a sample (with little accuracy) can provide the business executive with none of the definitive answers he may be seeking. Indeed, it is doomed from the start. On the other hand, too large a sample results in squandering money that could well be funneled into more productive channels.

A few additional words on cost are also in order. In the case of an interview-type sample, cost estimates can be readily obtained by multiplying (1) the number of interviews needed for required accuracy by (2) the cost per interview. There is generally no problem in estimating the latter. Indeed, if there has been previous experience in collecting this kind of information, a "rough and dirty" cost-per-unit estimate can probably be arrived at in a matter of minutes.

Once the total cost (unit cost multiplied by sample size) is calculated, the next question—and in some instances the most crucial of all—is, Can the firm afford it? In most cases there are limited funds available for sampling programs. And if the cost is so high that the program requires more money than the company is prepared to spend, the proposal may have to be scrapped or seriously curtailed. In any event, the time to find out about these cost limitations is early—before any money, time, or effort has been expended. The research junkyard is full of examples of projects scrapped in midstream because of cost overruns.

SIX

The Construction and Use
of Index Numbers

The so-called cost-of-living index is perhaps the most widely quoted statistic in the country. Housewives use it to bemoan increases in supermarket prices; workers use it to justify wage boosts; and armchair analysts cite it to blame one segment or another of the economy for today's inflationary woes.

But how many, aside from professional economists and statisticians, have ever stopped to take a closer look at this important measure—to see how it's constructed and to gauge what it can and cannot do?

Probably very few. Those who do, however, cannot help but be impressed with one key fact: It is not any monolithic measure, but rather a guide to the composite change in a large number of different commodities.

And therein lies its importance, for in most types of policy decisions in the business and economic sphere the crucial question is not so much how an individual price has changed but rather how the average price of many different items has changed.

This, in turn, raises another question: How can this average change be best computed? If all the commodities moved in proportion in the same direction at the same time, there would be no problem in effecting this kind of composite yardstick, for the percentage change in any one would be representative of the percentage change in all.

Unfortunately, the real world isn't that simple or orderly. Take 100 price or production items (indices cover many more areas than just price)—and the chances are that despite an overall uptrend or down-trend, there will be a goodly number of commodities moving in the opposite direction. And certainly not all would exhibit the same rate of change. The problem in this more normal type of case, of course, is to arrive at an average measure of change that might best describe the entire mix. This, in essence, is the function of an index number.

But the construction of such a yardstick is not without its problems. In production, for example, it is impossible to total the physical number of automobiles, refrigerators, TV sets, etc., in two different time periods and arrive at an overall average production change for, say, consumer durables—primarily because there is no physical unit of measurement common to all these items.

Consider for a moment the results of a simplistic approach where no attempt to find a common denominator was made. If there were 100 cars and 50 refrigerators produced in one period and 150 cars but no refrigerators in the second period, a simple adding technique would show 150 units produced in both periods—or no change. Yet it is obvious that the production effort needed to turn out 150 cars is considerably greater than the effort required to turn out 100 cars and 50 refrigerators.

In a similar vein, just as production of different items can't be added and then averaged to obtain a measure of overall output change, neither is it feasible to add and average reported prices to obtain a measure of the general price trend.

The index technique is designed to overcome this difficulty by convert-ing everything into relative or percent change. For it is possible to add and average two or more percent changes—provided, of course, that the relative importance or "weight" of each component is known. Put another way, the basic key to all index-number construction is the con-version of absolute numbers into relative terms.

There is another big plus to index numbers. Since everything is put in relative terms, comparisons are facilitated. Specifically, the index value for any given period is expressed as a percent of the value prevailing in a base or benchmark period. Convention has this base period labeled 100—with any reading above this signaling percent increase, and any reading below this signaling percent decline. Thus if the base year for prices is 1969 and the 1970 price index reads 104, it means that prices in 1970 were 104 percent of those prevailing in 1969—or more simply that prices rose 4 percent over the year.

The growing importance of the index approach is reflected in the ever-increasing number of statistical series now expressed in this form. The business section of almost any daily newspaper now carries regular indices on consumer prices, wholesale prices, industrial production, construction, stock market trends, productivity, unit labor costs, etc. In short, it is virtually impossible to assess today's complex business picture without recourse to index yardsticks.

Constructing an Index

The building of an index number is relatively simple—at least in principle. The discussion immediately below—involving the calculation of a price index—is designed to touch upon the major questions and steps involved.

1. Frequency. An index series can monitor prices on a daily, weekly, monthly, quarterly, or even yearly basis. The time span ultimately chosen should be a function of both price volatility and use. Other things being equal, if prices change frequently, the time period between calculations should be relatively short, and vice versa. On the use front, if you are using the index for gauging general inflationary pressures, then monthly calculations would probably suffice. On the other hand, purchasing agents who buy virtually every day would need statistical intelligence on a weekly or even daily basis.

2. Coverage. There is always the problem of which materials or items to include. While it is seldom necessary or even wise to include each and every item or transaction, the actual decision must depend in large part on the individual situation. Take a firm preparing a purchasing price index. One way to determine which materials to include might involve an ABC analysis of purchases. The A group would include the major cost items, and the B group the secondary components and materials, with the C items taking up the least part of the purchasing dollar.

Though it is desirable to include as large a share of the total material cost as possible, it is obvious that the more items there are, the greater the expense will be and the more calculations will be necessary. So A-group items with perhaps a sprinkling of B-group items might prove a satisfactory compromise. Sampling techniques can also help keep costs and calculations to manageable proportions. Also try to keep out minor items that might fluctuate widely in price—or items that might conceivably be eliminated in the near future because of product changes. A change in the mix of purchases will mean you have to revise the index.

3. Base period. There's always the question of choosing a base period against which to compare subsequent price changes. If a recession period (for example, 1960) were picked as a base, the index would have an inflationary tone since subsequent prices were on the uptrend. Similarly, a deflationary trend might be built into an index where the base period occurred at the peak of the business cycle. Sometimes it is preferable to use the average of two or three periods as a base; many government indices, for example, have generally been based on this latter type of approach.

4. Weighting. It is almost always necessary to assign to each item in the index its proper importance—or "weight," as it is called in the trade. Thus in a purchasing price index if the money spent on one key item amounted to 30 percent of total material costs, that item would get a weight of 0.30 in the index. Fix each weight accurately because these will determine the proportionate influence of each material in the composite index throughout the life of the index.

5. Raw-data preparation. Calculate the average unit price per period for each material involved. This is done simply by adding the total value of transactions in each period for each item—and then dividing by the volume (pounds, tons, etc.) represented by these values. Thus if you spent $45,000 in a period for 100,000 pounds of copper, the average price for the period would be 45 cents per pound.

6. Calculating item indices. In the base period each unit price is set equal to 100. For other periods the index for each material is determined by taking its average unit price in a given period and setting it over its unit price during the base period. Thus if copper were 40 cents per pound in the base period and 45 cents per pound in another period, the index would be $45/40 \times 100$ or 112.5.

7. Calculating the overall index. The index for each item determined in the preceding step is multiplied by its weight. All the weighted item indices for a given period are then added together to yield the composite index for the period in question.

Weighting Considerations

The previous discussion treated the problem of weighting in only very general terms. In actual practice, however, there are many different ways in which this key weighting task can be tackled. It is possible, for example, to use a single month, quarter, or year—or even a combination of years—as a basis for gauging individual item importance.

If the weights were to remain the same from year to year, then of course it would make no difference what period was chosen, for the results would be identical. But unfortunately, things just don't stand still in the workaday business world. So a decision on a weighting basis must always be made.

One of the most popular—and the one used by the government in computing its consumer and wholesale price indices—involves the use of base-period weights. In other words, weights are based on the relative importance of each commodity in the base period or the period set equal to 100.

But—as in all the other alternatives to be discussed below—this approach does have some shortcomings. It is possible, for example, that items that have been decreasing in price relative to others will attract an increased volume of purchases over time—simply because they are better buys. On the other hand, those items which have been increasing in price will tend to diminish in importance.

Take the hypothetical case where half the prices rise while the other half remain unchanged. This means that the price index (with constant base-year weights) will rise substantially. But buyers will tend to gravitate toward the items that have remained unchanged in price—if for no other reason than to hold down their costs. Thus, if diamond rings have gone up in price substantially, you might be tempted to substitute a relatively stably priced mink coat for your wife's birthday present.

In short, because of this substitution effect, any rise in overall procurement costs is likely to be considerably less than that indicated by the price-index advance.

A more commonplace example: If beef goes up in price but pork stays the same, the overall price index for meat will rise. But for those consumers who are willing to substitute pork for beef, living costs will remain substantially unchanged. It is for this reason that indices with base-year weights are said to have an upward bias. This type of weighting system is often referred to as *Laspeyre's formula* or *Laspeyre's index*—in honor of the man who first introduced the concept of base-year weights.

A second possible approach involves the use of given-year weights. Specifically, weights used are those associated with the year that is to be compared with the base year. But even bigger problems arise when this method of weighting is used.

For one thing, the computational work is greater because the weights change every year. Then, too, it is often impossible to obtain current-year weights without a substantial time lag. Still another problem: While each year is directly comparable to the base year (because the weights for both are the same), the comparison of any two given years is not (because each given year has a different set of weights).

By use of reasoning similar to that employed for the Laspeyre index described above, it can also be shown that this method gives undue weight to commodities that are declining in price. It is therefore said to have a *downward bias*. This second type of weighting system is referred to as the *Paasche formula*—again in honor of the man who devised it.

Combination weighting systems provide still other alternatives. One such approach (the *Marshall-Edgeworth formula*) averages given- and base-year weights. Another—the *Fisher "ideal" formula*—involves the computation of two different indices (one with given-year weights and the other with base-year weights). The two are then geometrically averaged to give an average result.

But the general criticisms leveled at indices with given-year weights apply equally well to these average weighting systems. For one thing, the weighting involves a different set of weights for each period—thus invalidating year-to-year comparisons. Also in most instances the current-year weights are not always immediately available—which delays release of up-to-date indices.

Another weighting problem involves what type of average to use when averaging weighted commodities. The arithmetic mean is generally used. However, in a few instances other measures of central tendency—particularly the geometric mean—are also employed. As a rule of thumb, the arithmetic mean should be used if the primary purpose of a price index is to measure the amount of money needed at different periods of time to purchase the same commodities. (This is, by the way, the most common usage of a price index.) On the other hand, the geometric mean is sometimes preferred when the index is used primarily to study the average behavior of price relatives.

Sometimes no weights are used at all in computing an index. This technique, known as an *unweighted* index, is used when either (1) the relative importance of each commodity is roughly equal or (2) the importance of each is not necessarily crucial for the purpose to which the index will be put.

On the latter score, the United States government releases weekly what it calls a "sensitive index" of spot commodity prices. The measure consists of just a few items that are sensitive to slight changes in supply and demand—and is used to measure shifts in general market sentiment. A rise means that demand is picking up—a decline, that demand is easing. Since the function of this measure is not to measure the average price of a market basket of goods, the weightings would add nothing to the value or accuracy of such an index.

Statistical Tests

Some analysts suggest mathematical tests to judge how good a particular price-index formula is. They are all based generally on the premise that an index representing the average of many items should behave the same way an individual item would behave over time. While few indices meet all the tests discussed below, this remains a goal that many shoot for. A brief description of some of the more popular tests follows:

1. Time-reversal test. This is used primarily to determine whether a given formula will work both ways in time—forward and backward. Specifically, if the base year and the given year are interchanged, the resulting values should be the reciprocals of each other.

This can best be illustrated by a simple example. If the price of an item goes from $1 in 1969 to $1.50 in 1970, we would say that prices rose 50 percent in 1970—or that the index for 1970 on a 1969 base was 150. On the other hand, if we compared 1969 relative to 1970, we would say that prices were 66⅔ percent of prices in 1970—or that the index was 66⅔ for 1969 on a 1970 base. Note that one figure is the reciprocal of the other—that is, the product of 150 and 66⅔ is equal to 100, or unity. When indices behave in this way, they are said to have passed the time-reversal test.

2. Factor-reversal test. This can best be explained by starting with the obvious fact that the value of any commodity is equal to its price times its quantity ($V = P \times Q$). Thus it follows that the ratio of value in, say, year 1 to value in year 0 would be equal to $P_1Q_1/P_0Q_0 = V_1/V_0$. Thus if both the price and the quantity of an individual product doubled in year 1, the value in year 1 would be four times the value in year 0. Or put another way, the value relative would be equal to the product of the price and quantity relative.

In the factor-reversal test, the same behavior is expected of price, quantity, and value indices. Specifically, if you have a quantity and a price index for year 1 relative to year 0, you would then expect the product of the two indices to equal the ratio of value in year 1 to value in year 0.

3. Circular test. This involves the shiftability of bases. Assume, for example, a shift from a 1947–1949 base to a 1957–1959 base. This, as discussed in the next section, involves dividing all the figures on the old 1947–1949 base by the 1957–1959 reading (also on the old base). The question then arises, Would the resulting index value—say, for 1969—obtained by this base-shifting process be equal to a 1969 index computed directly from a 1957–1959 base? If it would, the index is said to have passed the circular test.

Base-period Considerations

The choice of a base period was briefly alluded to in an earlier section. As pointed out at that time, a base period should be a typical or normal period against which change can be measured.

Generally speaking, it is desirable to shift the base period forward over time—primarily because as time elapses, the older base becomes less and less meaningful for analysis and interpretation. This is so for several basic reasons:

1. The index number on an old base becomes unwieldy over time. As the years wear on, price or production indices are likely to rise much above the 100 base-period level—making percent changes harder to evaluate. Thus a rise from 300 to 312 (which might occur say, after 10 years) is considerably more difficult to convert mentally into percent change than, say, a rise from 100 to 104. Also, the mind becomes accustomed to using 100 as a yardstick for measuring change—and finds it hard to adjust to any other benchmark, whether smaller or larger.

2. _The reliability of the average represented by the index becomes more and more questionable._   As times goes on, it is possible that the individual commodities being averaged will show wider and wider dispersion. This makes averaging—the raison d'être of index-number use—somewhat less meaningful. The average price of meat when pork and beef are fluctuating within 1 or 2 percentage points is a lot more meaningful for policy decisions than when 10 or 20 percentage points separate the individual products.

3. _Consumption patterns and tastes change._   What might be a typical market basket of goods purchased by the average wage earner 10 years ago is not a typical one today. Air conditioning was a luxury 10 years ago. Today for many it has become a necessity. Put another way, the weights or relative importance of individual items making up an index changes over time—and sooner or later the index must be adjusted to take these changes into account.

4. _The product mix tends to change._   New products come into the market as others are chased out. True, statisticians have devised a technical process for incorporating these changes, called _linking_. Unfortunately, this approach is not nearly as accurate as starting with a more up-to-date mix. Nevertheless, this is the way the government adds in new products and subtracts out old products from its existing indices.

5. _Basic quality changes over time._   The automobile turned out today is not the same product turned out a decade ago. Again statisticians have developed techniques for evaluating such quality changes and for correcting the index so that theoretically it will measure price changes of constant-quality goods. But the chances of error in such computations are much reduced when today's car is compared with the car turned out last year or the year before last.

For the combination of reasons listed above, the United States government tends to update its base about every 10 years. In 1971, for example, Uncle Sam went on a 1967 = 100 base after approximately a decade of 1957–1959 = 100. Most private indices also follow this same pattern—basically to facilitate comparison of their data with overall government yardsticks.

Finally, a few words on a do-it-yourself index-base shift are in order. Often, two series may be on different index bases—and it is desirable to compare the movements of one series with those of the other. In such cases it may be useful to shift both indices onto the same base. This is easily accomplished.

All that is required is to divide the index for each period by the value for the new period that is to be used as the base. For example, assume that on the old base the index reads 100 for 1968, 105 for 1969, and 110 for 1970. If we then want to convert to a 1969 base, we would simply divide each index number by the 1969 value of 105. Thus the new values would be approximately 95 for 1968 (100/105), 100 for 1969 (105/105), and 105 for 1970 (110/105).

Limitations and Problems

Some of the difficulties encountered in the construction of index numbers have been noted above in conjunction with the periodic need to update the base period and weighting systems. At this point a more detailed discussion of some of these problems—and of the other headaches involved in the construction and interpretation of index numbers—is in order. The discussion will be primarily in terms of price indices. But the same approach in most cases is equally applicable to production or any other series that lend themselves to the index-number approach.

1. Sampling. Generally speaking, any index number involves a series of sampling decisions. It is often physically impossible and generally of no practical value to include every price transaction. In any case, the usual approach for big government indices is to stress only the more important products, with just a sampling of the less important ones. Similarly, emphasis is usually placed on the most important sellers in the most important markets.

For example, if a limited number of commodities—say, 20—account for 90 percent of the universe or population, with an additional 1,000 products accounting for all the rest, then the 20 should be included with certainty. After that, some small number of the remaining items should be selected, primarily to determine whether the price behavior of the lesser items departs radically from that of the major ones. If, as is frequently the case, no serious differences exist, then including only the 20 items results in no great error in the index and at the same time yields a great savings in terms of time, effort, and money.

While no hard-and-fast rules can be established, it should also be noted that there are coverage limitations that are dictated by budget considerations. One can almost always squeeze a bit more accuracy out of an index. But the question is, Couldn't the money be spent more wisely in some other, more meaningful pursuit?

2. Statistical reliability. From the above discussion it follows that most indices are based on a sample of commodities which have been purposely selected rather than chosen by random methods. Therefore, the standard statistical techniques for evaluating the error in a sample are not applicable.

However, experience over a long period of time suggests that an index becomes increasingly reliable as the group of prices become larger. As the Bureau of Labor Statistics (BLS) puts it: "The reliability of a subgroup is greater than that of a product class, a group is more reliable than a subgroup, and the all-commodities index is more reliable than a group index."

3. Specification considerations. A cardinal rule of any price index is that the only variable that should affect the change in the index level is price. Therefore, it is highly desirable for prices used in the index to adhere to rigid specifications insofar as this is possible. Changes in physical characteristics, quantity discounts, credit terms, delivery charges, modes of packaging, and package size are examples of what might be considered specification changes. Thus when industry and trade practices shift, proper account should be taken of such changes by making appropriate statistical adjustments.

4. Quality changes. Clearly related to the specification problem is the one involving shifts in quality. To measure pure price change, one must deal with a product of constant quality. If quality deteriorates at constant price, for example, the consumer is obviously getting less for his money, and an index should rightfully consider this a price rise. Similarly, if one gets a better product at constant price, he is getting more for his dollar—or, in effect, a price reduction.

It is the task of the statistician to correct reported prices for such quality changes in order to avoid distortion. And indeed, the government does make such corrections. A case in point: Detroit's $78 increase in average auto price lists during the 1966 model year. The government statisticians, after examining the increase in great detail, however, felt that the new safety features introduced at that time more than compensated for the increase. In other words, the feeling was that the consumer was getting more car for each dollar spent. Thus the government rightfully decided that the car index that year should show a fractional decline.

But for all the pains the government takes, some quality improvements slip by unnoticed. One BLS official recently said, "It is likely that the incomplete representation of quality improvements imparts some upward bias (makes the index higher than it should be) in the long run."

However, keep in mind that quality can be a two-way street. For example, there seems to have been an evident decline in the quality of some services. Anyone who has attempted to have his car or TV set repaired can easily attest to that. Then, too, there is reason to think that in a period of intense prosperity, sellers tend to substitute higher price lines for low-end items at prices which reflect something more than the improvement of quality.

A more detailed discussion of the problem of quality and the specific techniques used for dealing with it is presented in Chapter 11.

5. Transaction prices.   More often than not, published or reported prices are nominal or list prices—serving merely as a basis for price negotiation. Nor can one rely on a constant adjustment in such prices to correct for what is usually an overstated price. The actual prices at which transactions are made will vary from time to time, depending on prevailing economic conditions, supply and demand factors, and the negotiating ability of the parties involved in the transaction.

It follows then—theoretically at least—that only transaction prices should be used. But this is more easily said than done. For one thing, buyers and sellers often prefer to keep their deals private. But more important—even if they were willing to disclose the correct price—the task of collecting the appropriate data would be enormous. Consider, for example, a steel price. It is a lot easier to go to a few sellers in this oligopolistic industry than to the literally thousands of buyers who might consume the given steel product.

The government from time to time has attempted to do some research or pilot studies on transaction prices—using buyer information. But it has always come up with the same conclusion: The cost and effort are prohibitive, and any money spent might be better utilized in some other statistical endeavor.

In essence, then, subjective judgment is the only way out of this impasse. When a buyers' market is known to exist, one can probably assume with reasonable certainty that official price indices overstate the existing price level. On the other hand, in times of shortage it is usually safe to assume that a price index is fairly accurate.

6. The changing product mix.   This problem was discussed briefly in the section on base periods. At that time it was pointed out that changes in product mix can best be handled by periodic changes in base years, at which time the basic weighting system is also revised. But this doesn't answer the question of what to do in the interim. It is certainly unrealistic to assume that the market basket of goods remains unchanged during the long periods of time between such periodic revisions.

Specifically, shifts in taste and technology are constantly taking place, so that any representative list of commodities, selected at any point in time, will with the passage of time reflect less and less the pattern so well represented by the list when it was first formulated. Because of this, additions and deletions must be made almost every year in the government's wholesale and consumer price indices. The frequency with which such changes are made will vary for different sectors of the economy and in different time periods, depending on how rapidly changes are taking place.

As noted above, linking is the statistical technique for making such changes. In the case where an old product has to be replaced in an index with a new one, the procedure goes something like this: The old item is dropped, and the new one is linked in at par, without affecting the level of the index. The assumption is that any difference in the price of the new article is commensurate with the difference in quality. Although the assumption is reasonable in competitive markets under normal conditions, this is concededly not a very sophisticated technique. It is the only available method, however, in the absence of detailed cost and quality analysis.

7. Absolute versus relative.   The typical price index is designed to measure change, not absolute levels of prices. This can best be appreciated if you consider the consumer price index for two different metropolitan areas. Specifically, city figures don't really show how prices in one area compare with those in another because they measure only price changes, not the actual level of prices. New York may have a higher price index than, say, Chicago, but that doesn't necessarily mean that prices are higher in New York than in Chicago. All it means is that prices have risen faster in New York than in Chicago since the base period.

8. Coverage.   Strictly speaking, the consumer price index is not a true measure of the general purchasing power of the dollar. It does not include prices of securities, real estate, construction, or a host of other things that a consumer can purchase. Even wholesale or primary market indexes, while a good approximation, are not a perfect measure since they're based on a relatively small sample of the many thousands of commodities that flow through these markets. Even assuming full coverage of all commodities, these wholesale indices still cannot be assumed to measure the average price paid by the corporate buyer since they exclude transportation, industrial services, and other purchases generally made on the wholesale level.

9. Definitional misunderstandings.   Up to World War II, the consumer price index was erroneously referred to as the cost-of-living index. Indeed, many people still cling to this name—forgetting that while the cost of living in any time period is affected by price changes, it is also influenced by changes in consumer expectations, which are constantly rising in a progressive society. When a person says that his cost of living is much higher now than it was 10 years ago, he may forget that he now expects and demands a more commodious home, a better car, and a more attractive wife. Perhaps, too, his income is higher and he is paying more taxes.

Keeping this in mind, it is important to remember that price indices are designed to measure change in a constant market basket of goods rather than upward shifts in living costs. Factors other than prices that affect the cost of living are best measured by government family-budget studies.

10. Specific requirements.   Despite the seemingly inexhaustible supply of statistical indices, industry and other private researchers frequently find it necessary to compile variants of government measures to meet special needs. Sometimes the government agency will be willing to do this for a fee. Thus, the BLS may be induced to develop a variant of a particular price index for, say, contract escalation. Or, if this isn't feasible, the private index maker, by using the basic government data, can do the recalculation on his own.

In any case, be prepared to lean heavily on government statistics—particularly when collecting price statistics on an industry-wide basis. That's because the Federal Trade Commission may rule that the intent of any such data collection is collusive and designed to fix prices illegally. So great is this restraint that no trade association will undertake to collect or publish prices from its members. Thus the price indexer is often forced to seek out government sources. Of course, the individual company is always free to construct price indices based on its own purchase or sales records. It is only when more than one company is involved in the collection that this collusion problem comes up.

Specific Applications

The number of purposes for which an index number can be used runs literally into the thousands. In today's complex, mathematically oriented society there isn't a business organization or government agency that can afford to ignore this vital statistical tool. Indeed, index numbers have become a virtual must in formulating policy, making decisions, and evaluating results. A rundown of some of the more typical broadgauge uses follows:

1. *A measure of progress.* Indices of output are needed to gauge real growth. Thus the Federal Reserve Board index of industrial production or any one of its hundreds of subgroups can be used both to quantify the advance in any one business sector—and to compare individual company performance with that of an entire industry. Many large, diversified firms today work up their own production index to facilitate this kind of analysis.

2. *A measure of inflation.* The rate of rise in any one of a number of widely watched price indices can be equated with the degree of inflationary pressure present at any given time. Government policy makers, for example, follow these price indices with particular attention before making fiscal and monetary decisions to either speed up or slow down the economy. For the business firm, analysis of price movements can aid in day-to-day buying strategy, suggest long-run procurement policy, and help pinpoint future cash flows.

3. *Deflating.* Price indices have become extremely useful tools for taking the price effect out of sales, inventories, new orders, and other key business parameters. Deflating permits comparison with physical volume measures such as production and provides a yardstick of real growth as contrasted to dollar growth. It is also often necessary for

preparing estimates of future capital equipment requirements. See Chapter 11 for a more complete discussion of the process of deflating and its various uses.

4. Seasonal adjustment. Indices of average movement to be expected over different periods of the year can take much of the guesswork out of estimating the meaning behind a jump or falloff in, say, sales or production. Called *seasonal indices,* these yardsticks in effect remove the seasonal influence, thereby facilitating evaluation of month-to-month changes that are due to the underlying business trend. Chapter 11 again provides a more detailed description of this particular index-number function.

5. Measuring productivity. By dividing a production index by an index of inputs, it is possible to derive a measure of production efficiency or productivity. A good example of this is the government index of output per man-hour. This is derived essentially by dividing the production index by an index of man-hours worked. These productivity figures, in turn, are then used for suggesting noninflationary guidelines for wage and price hikes. Other things being equal, wage hikes are thought to be noninflationary if they are equal to the rate of productivity advance.

6. Facilitating comparisons. If the GNP trend in Canada is to be compared with that in the United States, the easiest way to go about it is via index numbers placed on a comparable base. The index-number approach reduces both series to percent change—a prerequisite for comparison of two series of unequal magnitude.

7. Calculating farm parity. The determination of many farm prices is closely tied in with index numbers. Thus parity—a ratio of indices of farm prices received to farm costs—is used to determine United States support prices for wheat, corn, and a host of other major farm crops.

8. Cost-of-living escalation. Millions of workers today have collective bargaining agreements which call for automatic wage adjustments resulting from changes in the consumer price index. They are usually used to protect the worker against inflation. Thus if a union signs for a 4 percent annual wage boost in the hope of gaining a 4 percent hike in purchasing power, a cost-of-living escalator clause is just what the doctor ordered. If prices move up, say, 1 percent over the year, then 1 percent may be added to the worker's pay to maintain his real purchasing-power advance at a constant 4 percent level. There have even been instances when the consumer price index has been used to escalate alimony payments.

9. *Purchase-contract escalation.* One of the fastest-growing index-number functions is that of providing a measure to determine whether a previously signed contract should be adjusted to take into account rising costs. The remainder of this section will be devoted to this crucial dollars-and-cents use of price indices by businessmen.

The growing acceptance of this escalation technique is easily understandable, for it is often the only way a seller can protect himself in an era of constantly rising prices. A survey taken in 1962 indicated that nearly $14 billion in contracts involved such escalation. A conservative estimate today would probably put the total near $40 billion or even $50 billion.

The setting up of an escalator clause in a purchase contract, however, is a complicated affair. Generally, three major elements must be considered:

(a) INITIAL PRICE: All contracts must have an initial base price or rate at the time the agreement is set up. This is needed to provide a benchmark or frame of reference. Accuracy in setting up this base price is extremely important since escalation cannot correct an erroneous original price, and if the original price is incorrect, escalation will almost always tend to exaggerate the error.

(b) ESCALATING INDEX: All contracts must also have a price index that reflects changes in the cost of the commodity or service being escalated. The wholesale price index, or a component of this index, is usually used for adjusting prices of raw materials or production equipment. On the other hand, the cost-of-living index is generally used for escalating wages and items sold at retail levels.

(c) MECHANICS OF ESCALATION: Finally, the basic procedures for spelling out the escalation must be included in any contract. There are so many possible variations that every single step must be clearly defined. It's the only way to eliminate misunderstandings and reduce the need for haggling at some later date.

When selecting an index, it is important to be as specific as possible. The exact title and the index base period should always be indicated, and when the contract is based on the wholesale price index, it is important to specify whether the preliminary or final price index is to be used. That's because prices are often revised the month following the original month of publication.

In specially computed indices, the relative weights of each component should be specified. Take the example of an index for escalating the price of turbines where the agreement calls for an escalating index made up of several components. In such a case, each component must be clearly identified, along with its relative importance (weight).

Another escalation problem involves the stipulation of reference dates—the point-of-departure date for the escalating index and the dates for subsequent adjustments.

It's important not to confuse the point-of-departure date with the index base. If a contract were signed in June 1971, for example, the point of departure would be the index reading for June 1971—not the reading during its 1967 base period.

Reference dates for future escalation are also important. The contracting parties may prefer—for subsequent adjustment—to use a particular month's index, an annual average, or an average for three months. Any variation is permissible, but it must be clearly stated in the contract.

Still another factor to consider is the frequency of adjustment: If the buyer and seller agree that adjustments are to be made quarterly, the change in the product price can take place only four times a year.

Sometimes, however, a change in payment may be required whenever the index reaches a certain point—or when it changes by a specified amount. Thus if an increase of 1 cent were called for whenever the index moves up 0.5 percent, the time interval would be immaterial. (It might take one month or one year for the index to rise by that amount.)

There are no hard-and-fast rules on the frequency of adjustment. On the one hand, too frequent adjustments may create problems as a result of seasonal or erratic movements in prices. It is usually not advisable to make any changes unless they reflect basic changes in price trends. Use of quarterly, semiannual, or annual average indices will minimize such periodic fluctuations and result in smoother adjustment patterns.

On the other hand, in a period of continuous price movement in one direction, long intervals between adjustments may understate the true change. That's because escalator clauses adjust only for what has already happened. When prices are rising, payment increases lag behind index increases. And the converse is true during a period of falling prices.

The technical problem of choosing the actual type of price adjustment also tends to create problems. There are two basic methods of adjusting payments in accordance with price-index changes.

The first is the *percent-of-change technique,* which provides for applying some multiple of the percentage change in the base index to the base price. Thus a clause would read that prices will change a given percent for each 1 percent change in the index.

The second type is the *cents-to-point technique,* which provides that for each specified absolute change in the index, the price will change by some specified amount. Thus a clause might read that prices will change a given dollars-and-cents amount for each 1-point change in the index.

It's important not to mix the two up—for example, combining a point change in the index with a percent change in price. If used correctly, both methods give approximately the same results (see Figure 6-1).

PROBLEM--The XYZ Corp. purchases metal widgets at $2 per unit. It signs a long-term contract to buy them, with the proviso that the price can vary up or down with the composite price index for steel, which for the sake of simplicity is assumed to be 110.

The contract specifically states that the price of widgets will rise 1/2% for every 1% rise in the steel index. Subsequent to the signing the steel index rises to 120. What should the XYZ Corp. pay for the widgets?

SOLUTION--The increase in the steel index from 110 to 120 represents a 9.1% increase. Since widgets are to rise 1/2% for every 1% rise in the steel index, widgets would rise 4.55% for the actual recorded 9.1% rise in the steel index. This would boost the price of widgets to $2.091 (calculated by multiplying the base $2 price by 104.55%).

ALTERNATE--If the contract instead of using the percentage technique, had employed the cent-to-point approach, almost the same answer would have resulted. Specifically the contract could have read that the price of widgets would change 0.9 cent for every 1-point change in the steel index. Since the index went up 10 points, the price would go up 9 cents (0.9 cent X 10 index points). This would put the new price at $2.09, very close to the results obtained by the other method.

FIGURE 6-1 *Working out an escalation problem.*

It's also important to state whether escalation is to work only on the upside or whether it is to cover price declines as well as rises. Buyers generally should be interested in the two-way contract. This way they can get the benefits of lower prices, as well as give the vendor protection in case of rising costs.

Many variations on the two-way contract are possible. Some clauses specify that prices are to move down as well as up—but are not to drop below a specified minimum. For example, a product price escalation contract may call for a 1-cent decrease for every 0.4 index-point drop, but only down to an index level of, say, 98.0.

A final point to look out for is a possible correction in the index. The government sometimes revises prices going back six to twelve months. The contract should specify whether account is to be taken of such index corrections. In the same vein, there is often a major revision which goes back for several years. This is usually done to take care of new commodity specifications, new products, obsolete items, changing weights, and the need to update the reference base.

Usually such changes are taken into account by a statement to the effect that the agency issuing the price index will be the sole judge of the comparability of successive indices.

United States Price Indices

Today, with the voluntary cooperation of business and industry, the United States has the most comprehensive coverage of government-published commodity prices in the world. Indeed, one of the problems is that with so much pricing intelligence available, it is sometimes difficult to put it all into proper perspective. Generally speaking, there are many basic types of price measures available—each one designed to serve a different purpose.

More often than not, the price patterns and trends will differ index by index. This is because each index has a different composition and monitors a different part of the overall American market. For these reasons, it is important to know just what each index measures (as an aid to index users, the federal government bureaus that do the index computation work will provide detailed explanations of the indices concerned on request).

The following is a brief description of some of these different types of indices.

1. The consumer price index. Better known as the CPI, this monthly yardstick measures changes in retail prices of more than four hundred goods and services ranging from a can of soup to a new house—from a beauty parlor permanent to a funeral. The items chosen are representative of a typical market basket of purchases by city wage earners and clerical workers and their families—who make up about 40 percent of the United States population. (The CPI therefore isn't fully applicable to 60 percent of American consumers. Doctors, teachers, retired people, and others in that 60 percent do not buy the same things in the same proportions as the urban workers covered by the CPI. Indeed, the index is so poorly suited for farmers that the Agricultural Department publishes a separate retail price index for them.)

More specifically, the CPI covers prices of everything people pay money for in order to live—food, clothing, automobiles, homes, house furnishings, household supplies, fuel, drugs, recreational goods, doctors' and lawyers' services, beauty shop treatments, rent, repairs, transportation, public utilities, etc. It deals with prices actually charged to consumers, including sales and excise taxes. It also includes real estate taxes on owned homes, but excludes income and personal property taxes.

For the sake of convenience, the index is broken down into many different groups, subgroups, and individual commodities so that the prices of, say, food, meat, or even beef can all be easily compared at the same time. Similarly, consumer goods are split into durable and nondurable—and finally down into individual items such as automobiles. Similar breakdowns for services are also available.

Aside from determining wage rates for several million workers, the CPI is used as a guide to (1) general inflationary pressures, (2) purchasing power, (3) long-term rentals, and (4) pension payments. On the latter score, the payments made to more than one million retired military and civil service employees are affected by CPI changes.

2. The wholesale price index. This key yardstick of price pressures measures monthly changes at primary market levels—that is, the first important commercial transaction for each commodity. Most of the quotes are the selling prices of representative manufacturers or producers or the prices quoted on organized exchanges or markets.

Subindices are available for farm products, processed foods, and industrial commodities. Other useful breakdowns present these indices on (a) an end-use basis; (b) on the basis of the principal material component, such as rubber, metals, textiles, etc.; (c) by state of fabrication—raw, intermediate, and finished; and (*d*) on the basis of durability. In all, about twenty-five hundred separated index series, ranging from individual commodities on an eight-digit SIC classification up to the two-digit classification, are published.

The index has several important uses. First, as the measure of price movements at other than the retail level, it is one important indicator of change in the economy. The index is widely used by business and economic analysts in both setting policies and measuring their effectiveness.

Second, the index is used as a measure of the purchasing power of the dollar (excluding retail, where the CPI is used) and is therefore a key factor in the periodic adjustment or escalation of many long-term purchase or rental agreements.

Another important use of the index is by buyers and sellers of commodities—purchasing agents and sales managers. In most of these cases, it is not the total index but rather the group indices and the individual price series which are employed. Buyers of commodities are able to check both the amounts which they pay for goods and the general movement of their purchase prices against the index. The use of the index for checking absolute price levels is—as is generally the case for all price indices—sharply limited, however. The main goal has been to measure the direction and amount of change, and only incidentally to measure actual selling prices.

Wholesale prices as a measure of general and specific price trends are also widely used in budget making and review, both in government and in industry; in planning the cost of plant expansion programs; in appraising inventories; in establishing replacement costs; etc. The index is also used in LIFO (last-in, first-out) inventory costing by some organizations.

3. *The GNP price deflator.* This widely watched indicator is used to convert reported GNP into real dollars (dollars with inflation factored out). Thus this yardstick helps pinpoint that part of the GNP which is attributable to higher prices and that part which is caused by rising physical output.

One main plus of this measure is that it is all-inclusive, taking in wholesale prices, retail prices, service prices, etc. In short, it is designed to give a quick overview of price movement within the economy. It, too, is broken down into some major subgroups to help ferret out the cross-trends that invariably buffet an economy. On the other hand, some criticism has been leveled at this yardstick—mainly on the grounds that it has an upward bias and that it is published only quarterly, with a two-month time lag involved.

4. *The spot commodity index.*   This is probably the most up to date of all government indices. It is published weekly—with a lag of only a few days. However, it has only specialized use because of limited coverage. Also, it is the most sensitive and volatile of all the popular price yardsticks because it includes only those raw materials which are especially responsive to slight changes in market climate.

Historically, this index has been a fairly reliable early-warning barometer of broad-gauge shifts in the industrial price tone. But, because of its limited coverage (only 20 items or so) and because it is unweighted, analysts warn against its use as an indicator of overall price-level change.

5. *Diffusion indices.*   These are essentially combinations of other published indices. In essence, they give the percentage of individual price subgroups (which can be chosen as needed) that are rising and falling at any given moment. Put another way, a diffusion index is a simple summary measure which expresses, for a particular aggregate, the percentage of components rising over a given time span. It reflects only directions of change among the components, not magnitudes. Cyclical changes in diffusion indices tend to lead those of the corresponding aggregate indices.

SEVEN

Advanced Techniques

No doctor today would attempt to cure a complex ear or throat infection with a simple aspirin. Similarly, no statistician can hope to plan today's complex marketing strategies with the use of simple averages or index numbers. These tools are good as far as they go—but they don't go far enough.

In a sense they are only the primary building blocks—the base upon which to construct the highly sophisticated processing and analysis systems needed to run a business in the closing decades of the twentieth century.

One of the nation's top marketing executives sums it all up: "Everybody makes blunders—it's part of the game. The trouble is, however, that when the competition starts using scientific techniques, the game shifts in his favor. Conversely, if we can get the jump over our competitors, we then have the advantage."

The computer, of course, has furnished substantial impetus

to this new scientific approach. For without its ability to convert literally thousands of man-hours of manual work into a few minutes of machine time, little in the way of practical application would be possible. The computer's role in quantitative analysis will be discussed in greater detail in Chapter 9.

But computer or no computer, one point is clear: Statistical methodology today has advanced to the point where it can take much of the guesswork out of decision making.

Moreover, areas of application are no longer limited to the traditional marketing type of business problem. Over the past few years objective manipulation of data for better decision making has spread rapidly to such crucial areas as research and development, production, and distribution. Indeed, the statistical approach can even be applied to such seemingly complex and esoteric problems as long-range technological development.

In short, recent developments have served to build up the role of statistics in the modern business firm to the point where companies can no longer afford the luxury of the old-fashioned seat-of-the-pants approach. To be sure, the intuitive tack still has its place—but mostly as a commonsense check for some of today's more complex methodology.

But the new statistical revolution has created problems as well as benefits. For example, the tools required under the new quantitative approach require a growing amount of sophistication—in regard to both the techniques themselves and the way they are applied. But this needn't be any deterrent to the nonmathematically oriented executive, for it is still possible to achieve functional understanding without detailing the nitty-gritty calculation details.

The following discussion is aimed at providing management with the savvy and know-how to evaluate the most popular and useful of these new approaches—leaving the complex formulas and actual computational chores to the soldiers on the statistical firing line.

Time-series Analysis

This is essentially a method of forecasting the future on the basis of what is happening today and what has happened in the past. The name *time series* derives from the fact that historical information is analyzed on a temporal basis in an effort to discover any underlying or periodic patterns that would be of help in estimating future developments.

Business data are admirably suited for this type of approach because time sequence is almost always crucial when it comes to projecting changes in sales, production, costs, profits, etc. Indeed, a first approximation of tomorrow's results in these key areas can almost always be obtained by simple extrapolation of current and near-past performance.

To take a greatly oversimplified hypothetical example, if consumption of a commodity had been growing at, say, 5 percent in the late 1960s and 6 percent in the early 1970s, then this approach would put growth in the late 1970s at 7 percent.

While the concept is relatively simple, working out a time-series forecast can involve a lot of tough practical problems. Most of them are traceable to the fact that—aside from the basic long-run trend—company activity can fluctuate because of changes in general business conditions (the business cycle) and the time of year (the seasonal factor). Moreover, there are also irregular (nonpredictable) factors involved such as strikes, hedge buying, speculation, war scares, etc.

To impose some order on time series, statisticians generally divide all these influences into four basic categories: trend, cyclical, seasonal, and irregular.

1. Trend factors. The *secular trend,* as it is sometimes called, refers to the broad-sweep underlying movement over a period of several years, sometimes a decade or more. Thus, while sales of an item such as automobiles might level off or fall over a short period of time (perhaps because of a business depression), the overall trend over the long pull would essentially be up.

Figure 7-1 illustrates the long-term or trend approach to forecasting in the area of sales of domestically made automobiles. The first step is to collect yearly sales over an appreciable span of time. In this case, 18 years are used, and the information is plotted as shown in the chart.

Note that while the sales line sometimes zigzags, there is a general upward trend which seems to approximate a straight line. If a straight line were drawn through the data (as shown) and continued into the future, it would be possible to get an estimate of sales in, say, 1978 or 1980.

How can such a straight line be determined? Some analysts just draw in a freehand estimate and let it go at that. However, since two experts can disagree on the exact path of such a straight line, most forecasters use a mathematical formula which is supposed to yield the "best" line. Such a best line is calculated by the method of least squares, referred to also in the section on correlation below.

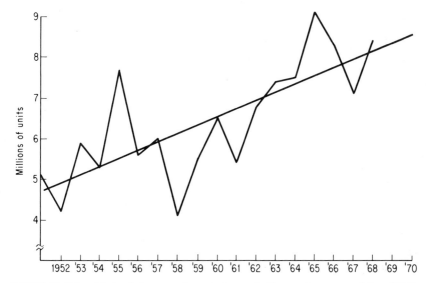

FIGURE 7-1. *United States sales of domestically made automobiles* (*U.S. Commerce Department*).

Once the equation for this best straight line has been determined (see Figure 7-1) it is a relatively simple matter to extrapolate and get the sales estimate for a future year. All that is required is substitution of the future year into the mathematical equation. The answer obtained is the sales forecast for the year in question.

In the example of automobile sales given above, substitution of, say, 1970 into the equation would yield a sales estimate of 8.5 million units. The actual results for that year were pretty close to the mark—8.6 million units. Forecasts, incidentally, may be obtained even more simply—just by reading the estimate for 1970 directly off the straight line in the chart.

Note, too, that the trend need not be linear. Indeed, in many cases it is not. As a matter of fact, there are many different types of growth trends or curves to choose from. Figure 7-2 shows some of the more popular ones—along with their mathematical formulas.

More often than not, for example, industry or company growth patterns tend to follow some variation of the S-type growth curve. An electric utility plotting its energy needs on the basis of population growth in a particular county found that the curve described just this type of pattern. The sales trend rose slowly at first as the area began to

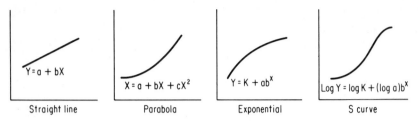

FIGURE 7-2 *Types of growth curves.*

develop. Then, as people rushed in, the rate of growth tended to accel-
erate. Finally, as the area became saturated, the growth pattern began
to slow down again.

2. Cyclical factors. But gauging the broad-sweep trend isn't
enough if forecasts cover only the next few years. That's because variation or change can be affected by cyclical and seasonal force. The former
are temporary waves of one to three years' duration, which are superimposed on the long-term trend. They can be caused by the business
cycle, or they might be inherent in the particular industry or product
being analyzed. Thus the textile industry tends to have a sales cycle
all to itself in which general business conditions play only a secondary
role in determining whether sales rise or fall in any given period.

Forecasting the cycle is generally difficult because even if the historical
cyclical pattern is isolated, the analyst cannot really be sure it will be
repeated exactly in the near future. Business cycles, for example, vary
in length and magnitude, and so do product-sale cycles—often because
of business-cycle variation, but many other times for totally different
reasons.

Nevertheless, many firms assume that these periodic undulations will
continue exactly as in the past. A growing number, however, are beginning to take a somewhat different tack where the general business
cycle is concerned. They note that economists are learning to cope with
these cyclical forces—and as a result this type of overall economic variation is becoming less marked.

To take this into account, their cyclical corrections to long-term trend
projections are slowly being reduced in magnitude.

Still others refuse to use any set cyclical correction formula. Instead,
they superimpose current government monetary and fiscal policies on
the existing state of the private economy—and come up with an estimate
of how much change, if any, is likely to occur in the current pace of

economic growth. This estimate is then factored into their estimates of company activity over the next year or so.

3. *Seasonal factors.* Even though the trend may be basically up or down and the broad, wavelike cycle is defined, there will still be other short-term (weeks or months) periodic changes depending upon the particular time of year. Since the seasonal influence is periodic and recurrent, it is a prime candidate for prediction—in the same way that the trend and the cyclical factors are predictable.

What's behind the seasonal factor? Apart from the repetitive effect of changing seasons, such things as customs, tradition, and other human factors also impose a seasonal pattern on economic activity. Behavior in social and business life often follows established usage: Holidays and religious festivals are established by law and tradition, and the calendar itself, which makes February 10 percent shorter than January, and April nearly 3 percent shorter than March, is a pervasive seasonal influence on economic time series. Thus conventional seasons are superimposed on climatic seasons to produce fluctuations that recur year after year.

The objective of seasonal analysis, then, is to determine the magnitude of this seasonal pattern—and then remove it from the data to determine the true, or underlying, trend. Statisticians and economists commonly refer to this as *seasonally adjusting* the data.

This can be accomplished rather easily—witness the spate of government and private reports that now come out regularly on a seasonally adjusted basis. Nevertheless, there is a tendency on the part of some businessmen to ignore such seasonal adjustment. Instead, they prefer to compare the data of any given month with those of the same month a year ago—thus canceling out any seasonal influence that might occur. But there is a serious weakness to this approach: It ignores the 11 months in between.

As one sales manager put it, "Any approach must start with knowing how the recent level and trend of sales compares with that of the immediate past as well as with that of a year ago."

Another sales executive tells this story of how his company erred seriously by using the year-to-year approach. The firm, which makes products tied to the construction industry, noted that its sales were running 5 percent ahead of the previous year's. Comparisons made several months earlier showed that a year-to-year gain was in the order of 10 percent.

The executives, after looking over the available data, were not too happy, but still felt that sales were in an uptrend. They drew up production schedules calling for a continuing increase in output compared with levels the year before. Much to their chagrin, they found that within three months they were actually running below the previous year—and they were stuck with a lot of costly inventory.

Hindsight revealed that if the data had been adjusted to take account of the seasonal pattern, the company would have noticed a change in trend (downward) before they gave the full-speed-ahead order. Use of seasonally adjusted data (data with the seasonal effect removed) could have saved a lot of red faces as well as a considerable sum of money.

A few words are also in order on the practical problems of computing the seasonal factor. Until recently it was a long, drawn-out affair in which the analyst first compared any given month's sales with a 12-month moving average (the average of sales six months before and six months after the month in question).

The theory here is that this 12-month moving average (since it encompasses a 12-month period) has the seasonal effect "washed out." Thus, the ratio of the actual value for a given month to the 12-month moving average gives the analyst a rough first approximation of the seasonal influence.

But since the cycle, trend, and random influences often tend to distort seasonal variations for any given year, the ratios for all Januarys, all Februarys, etc., are averaged together. The resulting average ratios are called *indices of seasonal variation*. This technique is generally known as the *ratio-to-moving-average method*.

All the more sophisticated deseasonalizing methods described below are based essentially on the principle described in this ratio-to-moving-average method. In a sense, they represent variations rather than any basically new approach.

One factor that has speeded the trend toward the use of seasonals in recent years is the development of advanced data processing techniques. With high-speed computers available, the seasonal adjustment problem can be solved with relatively little cost and effort.

The best-known computer deseasonalizing processes are the so-called census computer programs for isolating and identifying the seasonal component in sales and other time series. They are remarkably sophisticated programs, and the actual number of calculations is so large that the programs would be virtually impossible to tackle without a computer.

Before leaving the subject of seasonals, it should be pointed out that in some instances firms may be able to escape the computing chore. Specifically, they can "piggyback" on the seasonal calculations of others. Government statisticians, for example, have developed thousands of their own seasonals—many of which can be applied to individual company operations. In such cases all that is required is taking your company data and correcting it with these already available seasonal factors. Some practical examples of how this can be accomplished are given in Chapter 11.

4. *Irregular factors.* This is basically a catchall term for fluctuations which cannot be explained by trend, cyclical, or seasonal influences. In the past few years, some important work has been done in distinguishing between two types of irregular movements: random and nonrandom. This has given the analyst an additional aid in forecasting future trends.

Nonrandom irregulars cannot logically be identified as either cyclical or seasonal, but are associated with a known cause. They are particularly apt to occur in dealing with company data. An exceptionally large order will be received in one month. Sales in a particular month may be very large as a result of an intensive advertising campaign or an advance announcement of a forthcoming price increase. Where such is the case, a month or two of unusually low sales usually follow. It takes a moving average (the smoothing out of a given observation by averaging in observations on either side of the measurement in question) to smooth out irregularities of this sort.

Random irregulars are defined as all variations in a time series that cannot be otherwise identified as cyclical, trend, seasonal, or nonrandom irregular. These random factors are of short duration and of relatively small amplitude. Usually, if a random-irregular movement is upward one month, it will be downward the next month. This type of irregular can also be eliminated by a smoothing-out process such as the use of a relatively short-term (maybe three-month) moving average.

The major advantage of the overall time-series approach is that the computations can be accomplished with a minimum of experimentation.

Another plus factor, according to some proponents of this method, is that it doesn't have to rely on external or outside data. But whether outside factors are used or not, the approach is open to considerable criticism and danger if it is used alone, without offsetting alternative forecasts based on other techniques.

One reason for the danger of relying solely on the trend approach is that when the analyst extrapolates, he is assuming that what affected activity in the past will continue to affect activity—and in the same way—during future months and years. But this assumption is not always warranted. The introduction of a new product by a competitor, a faster rate of economic growth, a change in the foreign-competition picture, or a shift in consumer buying habits—these are but a few of the common changes that could make any straight statistical projection of past trends invalid and often dangerous.

Another problem, a technical one, stems from the fact that the trend method depends on an adequate number of past measurements or observations. For example, a new electronic device which has been on the market for only one or two years just wouldn't provide enough background for a trend, and if one were attempted, the results would be of highly questionable statistical significance.

Regression and Correlation Analysis

This is the branch of statistics concerned with measuring the relationship between two or more variables. Such a connection, if it can be established, can be extremely useful in making estimates and forecasts.

Before going into the actual techniques involved, it is important to distinguish between regression and correlation analysis. *Regression* defines the precise shape of the relationship (a straight line, a curve, etc.). *Correlation,* on the other hand, quantifies the degree of relationship (excellent, good, fair, poor, etc.).

A hypothetical example of how a builder of machine tools might use regression and correlation analysis can best describe how a relationship is set up and used as the basis for a sales forecast. Over a long period of time a company which accounts for a sizable portion of the nation's tool business noticed that its sales tended to follow the pattern of nationwide plant and equipment (P&E) purchases.

This relationship can be seen most clearly from the scatter diagram in Figure 7-3. The independent variable (the variable which effects the change—in this case P&E) is plotted on the horizontal axis. The dependent variable (the variable which is acted upon—in this case machine tool sales) is plotted on the vertical axis.

The essence of the regression-correlation technique is that it describes the type of interrelationship—in this case by a straight line, since the dots or scatter does seem to approximate a straight line. Once this

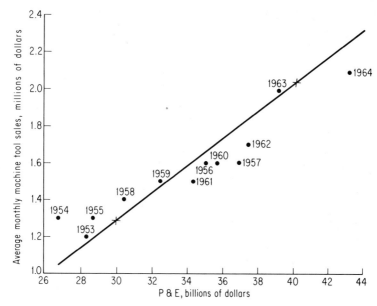

FIGURE 7-3 *Determining the best relationship.*

straight line is established, predicting becomes a relatively simple affair. Just take the estimate of P&E outlays (readily available from many sources, both private and public, for several years into the future) and read the estimated sales level right from the chart.

Before that can be done, however, the linear relationship must of course be calculated. One way is through a freehand estimation. But this sometimes creates difficulties. As in the case of the trend-line problem described earlier, two analysts may have entirely different opinions as to what is the right line.

Again the solution is to determine the one best line, based on the method of least squares—a description of which can be found in any elementary statistical text. This line is such that the squares of the deviations about it are a minimum.

The straight line is drawn on the chart by referring to the equation calculated by this least-squares method. In this case, sales $= -1.0 + 0.076$ P&E. By substituting in this equation any two values for P&E, we can obtain the best line which represents the relationship.

In the above example, let P&E equal $30 billion and $40 billion, respectively. The sales would come out to be $1.28 million and $2.04

million, respectively. These two points are marked with an "x" in Figure 7-3, are joined together, and become the best straight line.

Now the analyst is ready for the last step: the actual sales forecast. Assume, for example, that P&E purchases for 1971 are estimated at $44 billion. Go along the horizontal axis to $44 billion and then go up until the straight line is intercepted. The vertical reading of $2.33 million is the sales forecast based on the linear regression of sales on P&E.

To work mathematically rather than graphically, simply substitute $44 billion for P&E in the equation, and the same sales forecast is obtained.

The next task—once the forecast has been made—involves the problem of evaluation. This entails asking such key questions as the following.

1. *How good a relationship has been obtained?* Obviously, the more scatter there is about the line, the poorer the relationship and the less accurate the sales forecast. For this reason, it is important to know the degree of relationship.

With this in mind, statisticians have developed a measure called the *coefficient of correlation*. A value of 1 signifies perfect correlation (all the points lie squarely on the line). Lower and lower values for the coefficient of correlation signify less and less correlation (more and more scatter about the line). In the above case the coefficient of correlation comes out to .95—a relatively good relationship for the number of observations involved (12). Again the specific details on how to calculate the correlation coefficient are available in any elementary statistical text.

2. *How sensitive is the relationship?* Specifically, does a change in the independent variable evoke a big change in the dependent variable? This is important because even if the relationship is good, the sensitivity may be minuscule and thus hardly worth the trouble of working it out. The sensitivity is determined from the number in front of the independent variable—the number in front of P&E in the straight-line equation discussed above. It is usually called the *slope*. In the example under discussion, the slope (0.76) measures the change in average monthly tool sales per billion-dollar change in P&E. Thus $76 million of change in average monthly sales can be expected for every $1-billion change in P&E.

3. *Is there a temporal relationship?* Again looking at our illustration, it is conceivable that sales may lag a change in P&E by a few months or so. While the above example doesn't explore this possibility, such lead-lag relationships are extremely useful.

Thus, if sales did follow P&E by a quarter of a year, the analyst would not have to rely on a P&E forecast for his short-term (three-month) forecast. All he would do is plug in the current reading for P&E (assuming that the lead-lag relationship existed), and he would automatically get his sales forecast for three months hence.

4. Are all years applicable? Often in a scatter diagram one year or period lies far off the estimated line. If a special factor (known to the analyst) is responsible, this observation should then be omitted from the correlation.

A steel firm tells of its correlation of steel output with industrial production. It worked for all years except 1959, but there was a pretty good reason for the exception: nearly one-third of 1959 steel production was knocked out by a steel strike. The year was therefore omitted from the computation—and rightfully so, since correlation in this case was aimed at discovering the underlying relationship between steel and industrial production. Thus any observation that distorted this underlying relationship (such as a strike) would do more harm than good if included in the computation.

5. Should dollar or unit figures be used? Some firms prefer working their calculations out in terms of dollars; others do it in terms of physical units; and still others do it both ways. It's important to note that the underlying relationship may turn out to be somewhat different—depending upon the approach used. Often the choice is made in terms of what data are available and what the analyst wants and needs.

The thing to watch out for is the mixing up of terms. Thus, if one of the variables is in current dollars and the other is in unit or constant dollars, the correlation is likely to be distorted. It will, in general, be much less reliable than a correlation in which both variables (independent and dependent) are expressed in the same terms.

6. Would more than one predicting or independent variable improve the relationship? In many instances the addition of one or more other explaining or independent variables improves the relationship considerably. The general technique where more than one explaining variable is used is called *multiple correlation*. Thus, in the example above the average age of the current stock of machine tools might also influence the level of sales.

It should also be pointed out that the relationship need not be straight-line, or linear. It is possible, for example, to work out correlation problems using exponential, parabolic, or any other type of curve that can be expressed in mathematical terms. But also, as with trends, the computational volume increases with the complexity of the curve. This can be partially overcome by using logarithmic transformations.

Many firms report their best correlation results with a combination of several variables, all related in some kind of nonlinear relationship. This is usually referred to as *multiple curvilinear correlation*. As might be expected, however, the amount of arithmetic is prodigious and can usually be best accomplished with the help of a computer.

Another correlation problem often faced by analysts is the prediction of independent variables which, in turn, will be used to predict the dependent variable. Thus, in the machine tool example above, if no ready-made P&E forecast were available, one would have to be made before there could be any attempt to predict machine tool sales.

In short, there is little use in developing relationships with a series of data which are themselves so difficult to forecast that they do nothing to improve the reliability and validity of the basic forecast.

One last problem is that correlation presupposes an adequate number of observations to assure reliability. Thus, if sales on a yearly basis were to be predicted, at least seven to ten years of previous experience would be desirable. The technique would be useless if there had been only two or three years of experience because not enough observations would be available to establish a statistically valid historical relationship.

Step Regression

This is essentially a spin-off of a multiple correlation problem (where more than one independent variable are assumed to have an effect on the dependent variable). Stripped down to its basics, step regression is a computer-oriented approach which automatically (1) accepts those independent variables which have a significant effect on the dependent one and (2) rejects those which do not.

The computer is particularly important when there are 10, 20, or 30 variables that theoretically might be expected to play a role. That's because inspection alone can not usually reveal which are the significant variables. This fact can be discovered only through complex mathematical calculations—many of which could not be attempted without a computer.

Take a simple example involving the influence of income and education on the volume of purchases. Both may seem to play a role. But if income and education are interrelated (i.e., if high educations accompany high incomes), then it may be that the contribution of education, assuming a given income, may be negligible. In other words, income can explain variations in purchases almost as well as the combination of income and education.

The step-regression approach begins by selecting the independent variable that is most highly correlated with purchases—say, income. It then proceeds by incorporating education and all the other possible variables into the equation or dropping them; the decision to accept or reject is based on whether or not the variable in question adds significantly to the forecast. This procedure then continues until every significant factor is included in the estimating equation.

Model Building

Like correlation, this approach is concerned basically with discovering the influence of certain variables (independent variables) on other variables (dependent variables). It differs from simple regression and correlation in that many relationships are considered as part of the overall picture or model. Thus, if one were considering the launching of a new product, a model might be constructed consisting of equations or relationships describing the determinants of price, demand, and supply.

But more often than not, this approach tends to raise a knotty problem: The variables which are independent in one equation may be dependent in another. Thus while demand might determine price, it is also true that under certain conditions, the price of a product might determine its demand level.

In the example given above, the three equations describing the model might be expressed as follows:

▪ Price depends on (1) labor cost, (2) material cost, (3) overhead cost, (4) demand, and (5) supply.

▪ Demand depends on (1) price, (2) supply, (3) price of competing products, and (4) income.

■ Supply depends on (1) demand, (2) price, (3) labor cost, (4) material cost, (5) overhead cost, (6) material availability, and (7) labor availability.

At this point so many variables are related to so many other variables—each influencing the others—that the situation seems hopeless. But there is a way out.

Note, for example, that there are actually two types of independent variables in the above model or system of equations: those which are always independent and those which are never or only sometimes independent. Those which are always independent are classified as *exogenous variables*. Those which are not are classified as *endogenous*. Put another way, exogenous variables are determined outside the system; endogenous, within the system.

In our illustration the exogenous variables would be seven in number: labor cost, material cost, overhead cost, price of competing products, income, material availability, and labor availability. By complex algebraic manipulation it is possible to transform the three equations outlined above into a form in which each is expressed only in terms of exogenous variables.

All that remains, then, is to substitute values for the exogenous variables into the transformed equations—and come up with estimates of price, demand, and supply. Then, on the basis of this, a decision on whether or not to go ahead with the new project can be made.

Analysis of Variance

This is a technique used to decompose and analyze the different types of variation inherent in any business data. The aim is to see whether any particular source of the overall variation is statistically significant.

A typical application might be an analysis of the variation coming from several different machines on a production line. The approach assumes that the variation in the quality of products coming off any single machine is due to chance. This is then compared with the variation in product quality attributable to use of different machines. If this second type of variation significantly exceeds the chance variation inherent in the use of a single machine, it is assumed that there is a significant difference in the quality of the products turned out by the various machines.

This approach, which is an offshoot of sampling theory (see Chapter

5), is also widely used in advertising. A typical problem would involve measuring the effectiveness of an advertising campaign.

Exponential Smoothing

This is a relatively new technique of time-series forecasting which wrings the last ounce of information out of past company performance. However, it doesn't rely on blind repetition of previous cycles. Rather, through complex mathematics it relies on the use of a special type of moving average—one which gives increasing weight to nearby months and decreasing weight to months further back in the past.

Companies that use exponential smoothing cite the advantage that it doesn't rely on outside data. Thus it eliminates one of the drawbacks of techniques such as correlation, where reliable and timely outside data are almost always needed to predict future company performance.

Through the application of this one mathematical technique, it is possible to generate monthly raw data on sales, production volume, merchandise investment, employment, and profits. It is felt that the advantage of this technique lies in its greatly extended scope of practical application by virtue of its complete independence from outside indicators and personal judgment on the part of the forecaster.

PERT

This computer-based approach—short for Program Evaluation and Review Technique—is essentially a method involving planning, scheduling, and controlling the network of activities in large or complex projects. The basic problem here is that where a project involves a number of interrelated activities, they must be coordinated efficiently in order for the project to be completed satisfactorily.

Because of its ability to process huge quantities of data at great speed, a computerized PERT approach will quickly determine what activities in the project are critical—that is, which ones have to be completed within a specified time in order that the whole project may not be delayed. Equally important is the highlighting of activities that are not critical—activities that can be delayed without affecting the target date. This enables resources to be concentrated in the areas where they are needed most.

Such techniques of project control by computer are in widespread use in the building and civil engineering industries, where projects are usually large and involve many different types of labor and material.

Delphi Technique

Sometimes educated guesses figure importantly in statistical analysis. The so-called Delphi technique (a systematic utilization of the intuitive judgment of groups of experts) is a case in point. It's based on the premise that when nobody has perfect knowledge of the future, many heads are better than one or a few.

The technique usually works something like this: First, a panel of experts on the particular problem at hand is selected from both inside and outside the company or organization ranks. Each expert is asked to make forecasts anonymously. For example, panelists in one Rand Corp. experiment were asked to say by which of 30 breakthrough dates they estimated that 20 percent of the world's food would come from ocean farming. (Half thought this point would be reached before the year 2000, and the other half, after.)

Each panelist then gets a composite feedback of the way the other panel members answered the questions, and a second round of forecasting begins. The process may be repeated two, three, four, or more times. Since the identities are kept anonymous, a panelist can more easily change his mind after reading other opinions. In a committee session, he might care more about defending his original idea than about coming up with good predictions.

What Delphi does, say proponents of the technique, is to refine the judgments of experts. The first time around, definitive answers are not expected. Successive iterations improve individual answers, and better individual responses add up to a better group judgment.

The pioneer in this approach (The Rand Corp.) also has an answer for skeptics who find Delphi interesting but argue that they can make just as good a prediction with a Ouija board. In a series of controlled experiments, Rand asked a panel of nonexperts to use Delphi to come up with answers which could easily be verified in an almanac or company records. Sample questions were: How many popular votes were cast for Lincoln when he first ran for the Presidency? What was the average price a United States farmer received for a bushel of apples in 1940? In most cases, the estimates were within the ball park, and the results showed that Delphi's feedback techniques sharply improved the answers in successive forecasting rounds.

In some ways the Delphi technique is similar to another pragmatic approach to quantitative analysis known as *Bayesian statistics*. For it, too, is a new and controversial method for dealing with future uncertainties. Basically, it incorporates the firm's own guesses as data in the calculation of forecast. It has opened the way for a quantification of many marketing problems. On the other hand, conservative statisticians have attacked the Bayesian method, calling it nothing more than a "quantification of error."

Linear Programming

Within the past few years linear programming has become one of the most promising tools of scientific management. Essentially, LP is a technique for determining how to employ limited capacities or resources to the best advantage—with the best advantage defined as least cost, highest profit, greatest quantity, etc. Put another way, LP provides a systematic approach for using data and facts to arrive at decisions which were previously reached on the basis of intuitive judgment.

The mathematics are too complex to be explained in this kind of "how-to" book. More important is the need for management to be aware of the many types of problems to which LP can be applied. One very common application involves the allocation of work on existing machines. LP loading can generally (1) reduce work-in-process inventory, (2) permit meeting of delivery schedules, and (3), most important of all, increase the effective capacity of a given plant.

The classical transportation problem is another type of question that can be answered through the LP technique. In essence this involves determining the lowest transportation or distribution cost of a program for shipping material between various company origins (warehouses, factories, mines) and various destinations (customer plants). It involves mathematical manipulation of capacity at origins, customer requirements, and transportation rates and distances.

Game Theory

This involves the investigation of rational decision making in the context of uncertainty and/or incomplete knowledge concerning the moves of competitors. While the mathematics (based on probability) are a bit complex, the procedures can easily be routinized. Consider, for example, the following simple illustration.

Two competing companies—Company A and Company B—want to bid on a contract. Here is their situation: If A bids and B does not, A will lose $20,000. If both bid, A thinks he can gain $10,000. If neither bids, A thinks he's $30,000 better off than B. If B bids and A does not, A figures he's $20,000 ahead. The solution would follow along these lines:

STEP 1: A must sort the possible repercussions of his moves into two logical categories: (a) what happens if he bids and (b) what happens if he does not bid. This would normally be done via a box or matrix as follows:

	B Doesn't	B Bids
A Bids	− $20,000	+ $10,000
A Doesn't	+ $30,000	+ $20,000

STEP 2: Find the lowest value in each horizontal series and the highest value in each vertical series. Here's what you get:

	B Doesn't	B Bids	
A Bids	− $20,000	+ $10,000	− $20,000
A Doesn't	+ $30,000	+ $20,000	$20,000*
	$30,000	$20,000*	

STEP 3: Note that the figure $20,000—marked with an asterisk for accent—is common to both the horizontal and vertical series. The correct action for A, then, is the one suggested by the $20,000 figure; that is, he should not bid, and B should.

Very seldom, however, does a common number emerge from the matrix. In most cases the solution is more complex. Changing the numbers in the matrix and simplifying them for illustrative purposes gives the following results:

	B	
A	4	8
	5	3

If you follow the procedure in the game played above, you'll find that no common number emerges from the new matrix (results are 4 and 3 in the horizontal series and 5 and 8 in the vertical series). That's a tip-off that no single strategy is going to suffice in this more complex situation. Each player will have to mix his actions. Here's how it's done:

STEP 1: In both the horizontal and vertical series subtract the smaller from the larger number as follows:

B

4	8	4
5	3	2

A

1   5

STEP 2: Reverse the position of the numbers on each side as follows:

B

4	8	2
5	3	4

A

5   1

STEP 3: The answer to the strategy to be followed is now available in terms of odds. A should bid two times and refrain from bidding four times out of every six plays. B, on the other hand, should refrain from bidding five times and should bid once in every six opportunities. A few additional points to remember:

1. The above is geared to playing a continuous game—one that's repeated over and over. But what if you play the game only once? Thus if you are A with a two-bid–four-no-bid mix, arrange for a chance drawing where the odds are two to four in favor of making a bid. Then follow the strategy indicated by the draw.

2. In subtracting a smaller number from a larger number, remember your algebraic rules. A minus sign becomes a plus in subtraction. Thus 4 minus the quantity −3 yields an answer of +7.

The above, of course, are very simple games. But whether simple or complex, they are playing an increasing role in business strategy. To be sure, there are limits. Games are not the cure-all that some proponents of the approach would claim them to be. But neither are they the idle toys that some critics say they are. Like all statistical techniques, they are tools providing a means to an end.

Simulation

This is essentially a hypothetical testing, in contrast to field testing of the consequences of alternative business decisions. A company might want to use this technique to check, say, the effect of a price change on sales before actually going through with any real price change. The name *Monte Carlo simulation* is applied to simulation in which the effect of one variable or another is quantified in probability terms.

Many of these new techniques such as linear programming, simulation, and game theory are usually lumped together under the term *operations research,* which is defined simply as the application of sophisticated mathematical techniques to business decision making. While these OR programs can't mastermind corporate strategy, they can extract the last ounce of value out of every piece of objective datum the company may have at its disposal.

As one practitioner summed it up: "The new approach is nothing more than breaking down a problem into its basic elements and then looking at all the possible alternatives—not just the obvious ones."

EIGHT

Advanced Applications

The use of the sophisticated techniques discussed in the previous chapter is only one element—albeit a significant one—in the business statistician's stable of analytic tools. Equally important are the practical applications which stem not so much from direct use of established techniques as from manipulation and custom-tailoring to meet specific analytic needs.

The four areas discussed below, for example, stress some of the imaginative and practical applications of basic theory available to a skilled corporate analyst. Take the first one: statistical quality control. From a mathematical point of view, anybody who understands the basic techniques of statistical sampling can handle this important subject—for statistical quality control is little more than applied sampling theory. The only difference is that the required techniques are mathematically shaped to fit a particular business need—in this case, ascertaining whether or not quality is up to par.

At other times very useful applications can be derived from relatively simple principles. Thus the discussion on inventory control below is based on the commonsense idea that the more frequently one orders, the higher the order cost but the lower the inventory carrying cost. Similarly, the less frequent the ordering, the lower the order cost but the higher the inventory carrying cost. The problem then revolves around finding a formula that will yield an order and inventory strategy that will minimize overall costs.

Also derived from a relatively simple idea is the rather complex section on how to make a capital investment decision. It has its basis in the simple discount idea that a dollar earned five years from now is worth less than a dollar earned today. To be sure, the ramifications are complex and far-reaching. But without the basic idea of discounting, the application could never have been developed.

Finally, the last section on input-output is grounded on nothing more than the idea that one firm's purchase is another firm's sale—plus the fact that a purchase in one industry sets off a chain of reactions leading to further purchases in a wide range of other industries. Here, too, the ramifications and repercussions are seemingly limitless, requiring substantial computer usage. But again the basic principle is simple; only the application is difficult.

Only four examples have been chosen here—primarily because of space limitations and the fact that these are important ones and are likely to prove useful to the majority of firms. The actual number of possible sophisticated statistical applications to specific business problems is virtually infinite—limited only by the imagination and resources of the individual firm.

Before going into the detailed discussions, it should also be pointed out that application is a key factor in the solution of almost any type of business problem. Techniques provide only the starting point. The actual details are all functions of individual company needs, which are in turn functions of cost, type of industry, type of firm, size of firm, etc.

Thus, a sales forecasting system which works perfectly in one firm may turn out to be inadequate or even a dismal flop in another. There are just too many variables in today's complex business world to permit any pat, across-the-board solution that applies to each and every firm.

As one statistical analyst recently put it, "Forecasting may show you where you're going and give you a map for the route—but you still have to do the driving."

Quality Control

The ever-present problem of maintaining and upgrading quality has led in recent years to increasing reliance on statistical quality control, a technique aimed at measuring variation in production and ultimately at keeping such variation within tolerance limits prescribed by buyer needs.

Surprisingly enough, SQC does more than just cut costs by pointing out areas where defective output is centered. It also saves by telling when more quality is being supplied than is really needed. Thus if quality is far better than the engineering tolerances call for, it may be possible to reduce costs by using different materials, by cutting down on inspection, etc.

SQC need not be limited to complex operations. Manufacturers of any mass-produced product can use this technique to advantage. Indeed, some of the most conspicuous examples of savings are in relation to comparatively simple products.

What SQC is can best be defined by describing what it aims to do. Its goal is to provide answers to these two basic questions: (1) When should a manufacturing process be left alone, and when should corrective action be taken? (2) What should the buyer and seller do with an individual shipment that is submitted for acceptance?

As to the first question—that of determining whether corrective action is needed or not—SQC resorts to one basic technique: the control chart. This is essentially a graphic procedure for distinguishing between quality variation that is normal (due to chance) and quality variation that is excessive (over and above what might be expected on the basis of mathematical probability theory).

For example, if you tossed a coin 100 times and got 100 heads, you would suspect that the coin was loaded ("out of control," to use SQC jargon). Such a result is pretty far removed from chance factors, which predict that approximately 50 heads and 50 tails would be obtained in 100 tosses.

The control chart (see Figure 8-1) essentially sets the limits, or cutoff points, beyond which chance variation is extremely unlikely. If quality falls beyond these points, the manufacturing process is probably out of control, and some significant factor (poor labor, machine wear, etc.) is acting to cause too much variation—too much variation in the sense that it cannot be ascribed to chance.

Another advantage of the control chart is that it can be used for both measurable data (specific dimensions) and attribute-type data (good or bad, go or no-go gauges). The example below deals with measurements, but an equally simple chart could have been used for attributes.

It should be noted that the chart can also point out potential danger spots—even before they actually start hurting. Thus when all the points lump close to the control limits for a period of time (even though still technically under control), it could mean a shift in the manufacturing process level. It's always desirable to look into this before things actually go out of control.

Another factor to remember: Even though a process may be in perfect statistical control, many items still may not be meeting engineering specification tolerances. For example, your engineering tolerances might be so tight that they don't account for normal variability. In such a case the process is under statistical control but not engineering control. Both are important, and both should be evaluated.

One quality control engineer likens the control chart to a "getting smart" approach as contrasted to a "getting tough" approach. It will do the quality engineer no good to start searching out causes of variation beyond engineering tolerances when such variation is due to chance. He could blame it on the foreman, the worker, or the machine; he could even fire the whole staff. The results would always be the same: The looked-for improvement would never materialize. The point is to exert pressure—but only when pressure will pay off.

The following example illustrates how an actual control chart might be set up: The ABC Pipe Company had been receiving a growing number of complaints from buyers. Their beef: A substantial percentage of pipe fittings, which require very close tolerances, had been shipped with too much variation in width. All this—in addition to creating buyer annoyance—had resulted in considerable expense for ABC, for the firm had to absorb the cost of return, of rework, and of reshipment back to customers' plants.

ABC decided to take the bull by the horns and set up a quality control program. The aim was first to get an early-warning system to signal when width variation was becoming excessive and second to suggest reasons as to why their output was going off the beam.

A quality control engineer was called in, and as a first step he recommended the setting up of a statistical control chart. He had the foreman take a sample of five fittings off the production line every hour. After 30 such samples were taken, the data were turned over to the quality control engineer, who then proceeded to work up a control chart as shown in Figure 8-1.

The plotted points on the chart represent the average fitting width for each sample of five—with the first dot representing the average width of the first sample, the second dot representing the average width of the second sample, etc.

The center or average (solid) line represents the grand average of all sampled fitting widths (the average of all the sample averages). This is assumed to stand for the typical or normal width.

The outlying (dashed) lines stand for the control limits. When any given sample average falls outside either of these two lines, it indicates that width variation is due to some assignable cause—and not to pure chance. These control limits are relatively easily calculated on the basis of probability theory (the laws of chance).

A quick glance reveals that lack of control occurs at samples 7, 8,

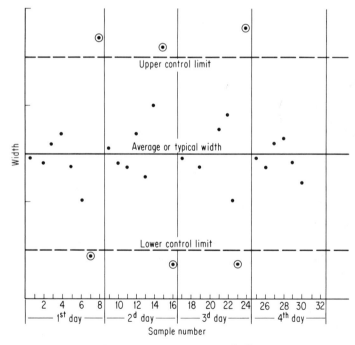

FIGURE 8-1   *The ABC Company control chart.*

15, 16, 23, and 24 (circled dots on the chart). Further examination reveals that these samples were taken during the seventh and eighth working hours of each day—which suggests that production workers might have been fatigued and hence were being less exacting in making each individual fitting.

ABC, with this possibility in mind, instituted a coffee break every afternoon at 3 P.M. And it paid off. Subsequent production runs remained under control, with little sign of systematic deterioration in quality during the late-afternoon hours.

In fact, accuracy improved so much as to result in better-than-needed buyer tolerances. It eventually prompted ABC to reanalyze its whole production process. The company found that it could cut down on its production costs at the expense of only a slight increase in width variation—not enough to interfere with buyer specifications.

The resulting process change was then instituted, and the cost savings were shared by both the ABC Pipe Company and all its customers.

A few words about a second major area of SQC are also in order: What should the buyer do with an individual shipment submitted for acceptance? Actually, the buyer has three basic alternatives: (1) take the shipment on faith without any inspection, (2) proceed with 100 percent inspection if it is nondestructive, or (3) take a sample.

It is obvious that the first approach is dangerous and can lead to severe production bottlenecks and heavy expense if the merchandise is faulty. The second alternative—100 percent inspection—has the disadvantage of being very costly and time-consuming.

Moreover, there is a fallacy connected with 100 percent inspection. It has been found through experience that the effect of inspection fatigue often results in an efficiency level of only 85 to 90 percent. This means that on the average, only about 85 to 90 percent of defective pieces in a lot might be discovered by 100 percent inspection.

The third alternative—known as *acceptance sampling*—is a middle-of-the-road compromise aimed at overcoming the shortcomings of blind acceptance on the one hand and 100 percent inspection on the other. Like the control chart, it uses graphic techniques based on mathematical probabilities to determine whether to accept or reject a given shipment.

Acceptance sampling has an added advantage: Both buyer and seller know just what they're getting and what they're risking. However, if no mistakes can be tolerated—either accepting a bad lot (important from the buyer's point of view) or rejecting a good one (important from the seller's point of view)—then acceptance sampling is not the right alternative to choose.

The technique for acceptance sampling can again be demonstrated by means of charts called *operating characteristic* (OC) curves. In essence, they give the risk involved in any particular sampling plan (Figure 8-2). Again a simple problem might best illustrate how the OC curve works. The XYZ Machine Co. purchases widgets from the Acme Widget Co. in lots of 1,000. Full 100 percent testing is too expensive, and yet the firm can't afford the chance of working with too many defective widgets. (From XYZ's viewpoint, a lot with 2 percent defective is a bad lot.)

Statistical quality control analysts of both firms agree on a sampling plan to avoid acceptance of bad lots. The details: 100 widgets will be tested out of each lot, and a lot will be accepted only if all 100 sample widgets are OK. If there are any defective pieces in the sample, the whole lot will be rejected.

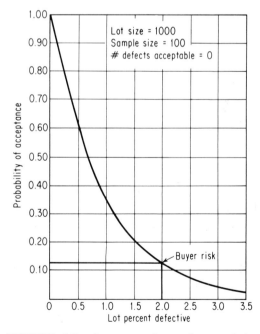

FIGURE 8-2 *An   operating   characteristic curve.*

QUESTION: What risks are both buyer and seller assuming in agreeing to this sampling plan, and are the risks equitable?

ANSWER: This can be read off an OC curve as shown in Figure 8-2. This gives a graphic picture of what the buyer and seller can expect the sampling plan to do to both bad and good lots over the long pull.

The curve is unique in that it applies only to situations in which there are lots of 1,000, samples of 100, and no allowable defects. For every different combination of sampling specifications there would be a different OC curve.

As for reading the chart, the horizontal axis represents varying degrees of lot quality expressed in terms of percent defective. And the vertical scale gives the probability that if a lot of specific percent defective is submitted, the sampling plan will accept the lot.

Then, referring to the chart, if a bad lot (2 percent defective) is submitted, the probability of accepting it is about 12 percent. These results are obtained by (1) locating 2 percent defective on the horizontal axis, (2) going up until the OC curve is reached, and (3) reading off on the vertical axis the probability of acceptance.

Thus the buyer will accept a bad lot (2 percent defective) about 12 percent of the time. This is the risk he is taking. But, as the chart indicates, if lots of poorer quality are submitted, the buyer's risk will be considerably less.

Now for the seller's risk. Assume for a moment that the seller submits a lot with only 0.5 percent defective (a very good lot). Under the sampling plan, the probability of a buyer's accepting it is about 60 percent. Or put another way, the probability of the buyer's rejecting this good lot is 40 percent. This is the seller's risk—the risk of having a good lot rejected.

It might seem to a buyer that he should be concerned only about keeping his own risk low. But this is not necessarily true, for if the seller is forced to reject a considerable portion of his good lots, his costs will rise, and this will ultimately be reflected in the price of his widgets.

The aim should be to hit on a sampling plan that is acceptable to both buyer and seller. It's possible, for example, to give the buyer the same protection and yet reduce the seller's risk—when the sample size is increased and the criteria of acceptance are relaxed (the lot will be accepted when the sample shows as high as one or two defective widgets).

The payoffs from the SQC approach (control charts and acceptance sampling) are many:

1. Better quality: A point plotted beyond the control limits is like a warning bell that something is amiss and needs looking into.

2. Lower production costs: Because production is stopped when the process shows signs of going out of control, there is less chance for a pileup of rejects and hence less rework, less scrap, and often much lower overall labor costs.

3. Reduction of bottlenecks: Since production is more reliable, there is less likelihood of a tie-up on production lines due to faulty materials or components.

4. Reduced inspection costs: If a product is known to be under statistical control, there is less need for sorting and 100 percent inspection.

5. Increased sales revenue: With better overall production assured, the firm isn't faced with the sales-reducing factors of (a) lower prices for downgraded merchandise, (b) settlement of customer claims, and (c) charges against the guarantee account.

6. Engineering guideposts: If a process is in control but not meeting engineering (as opposed to statistical) tolerances, then SQC suggests three alternatives: fundamental changes in the process, changes in the specifications, or 100 percent inspection.

7. Vendor rating guideposts: Since SQC is a quantitative rather than a qualitative approach, it can help the buyer set up numeral vendor rating systems.

8. Better buyer-vendor relationships: Since SQC is an impartial statistical approach, once buyer and seller agree, there is little chance of any bitter hassle over quality performance.

Inventory Control

With inventory carrying costs now running upward of 24 percent a year, keeping a tight rein on stocks has become a major goal for most well-run firms. That these firms have succeeded can be seen in the declining inventory/sales ratios throughout the economy. Thus, in the past two decades the key purchased-material-stock/sales ratio for durable goods has declined from a high of 0.71 to a low of nearly 0.57—a drop of nearly 20 percent.

To be sure, better management control, with an able assist from the computer, has helped pare unnecessary stock. But a good part of this development has also been due to increasing usage of mathematically oriented inventory formulas which suggest the most economic order value (EOV) or economic order quantity (EOQ).

Behind the EOV-EOQ concept is this commonsense thinking: If you cut down on the number of orders you place, you'll save labor and paper work, but if by doing so you increase their individual size to the point where you tie up too much money and storage space, you'll erase—or exceed—your gains. EOV tells you exactly where the true balance lies.

What has made this subject so difficult heretofore is the computation of the balance. The mathematics involved in it are time-consuming and require painstaking effort but they are not nearly as complex as commonly supposed.

Basically it boils down to this simple formula:

$$\text{EOV} = \sqrt{\frac{2RC}{A}}$$

where R = cost of placing one additional order (marginal order cost)
C = dollar usage per month
A = carrying costs per month expressed as a decimal

Next assume a simple problem involving the inventorying of steel by an appliance maker where (1) the average monthly usage of the steel is \$10,000; (2) the average carrying costs are 24 percent per year, or 2 percent per month; and (3) the order cost is about \$4. To determine the EOV, simply substitute in the formula given above:

$$\sqrt{\frac{(2)(4)(10,000)}{0.02}} = \sqrt{4,000,000} = \$2,000$$

When setting up an EOV system, however, it is best to work systematically, following the basic steps.

1. *Choose the appropriate items for control.* Your EOV program will work best if you concentrate on items that (*a*) consume the bulk of your expenditures and (*b*) have a high turnover. On the other hand, forget about such purchases as custom-made products, capital equipment, or items bought on tricky world markets.

In general, choose products (*a*) whose annual use can be predicted fairly well, (*b*) whose price is reasonably stable, (*c*) whose shelf life is longer than the inventory cycle, and (d) whose chance of obsolescence is small. Such a list should include most raw materials, standardized components, subassemblies, industrial supplies, stationery and office supplies, and spare parts.

2. Figure your order cost. For EOV purposes, order cost is *not* based on the expense of running your entire department (which obviously would stay in existence whether you used EOV or not).

Order cost here applies only to extra cost—specifically, that extra expense (mainly for labor and paper work) which you incur when you place one more order. Economists call this concept *marginal cost.* You must have marginal cost data to use this approach.

To get these data, first have your accounting department supply (or estimate) information about costs of salaries and fringe benefits, telephone, telegraph, travel, expense accounts, printing stationery, postage, rent, etc. These costs should be totaled for all interested departments including purchasing, receiving, stores, inspection, accounts payable, requisitioning, etc. Armed with this information, you now can determine your marginal order cost in one of three ways:

(a) SIMPLE METHOD: Suppose that your total purchasing department cost was $10,000 and that 1,000 orders were placed. You estimate that if you bought the same volume but used only 800 orders, the total cost would be $9,000—a difference of $1,000, or $5 per order. Similarly, you estimate that reductions in other (nonpurchasing) departments would come to $3 per order. Your marginal order cost in this case is $8 per order.

(b) SOPHISTICATED METHOD: If your accounting department keeps extremely detailed records, it can estimate the extra factor in each item noted above and give you a per-order total.

(c) COMPARISON METHOD: If yours is a small or medium-sized manufacturing firm without detailed records, pick a low order cost, say, $2 or $3. Also pick a low figure if you are in a relatively low-profit industry—metalworking, textiles, or food processing, for example.

Pick a high figure (say, $6) if yours is a larger firm or is in a comparatively high-profit industry—electronics or chemicals, for example.

3. Figure your carrying charges. Here again, you want only extra (or marginal) charges. Typically, such marginal charges range from 24 to 36 percent of inventory cost per year (or 2 to 3 percent per month).

Table 8-1 is a composite of recent estimates made by a number of

TABLE 8-1

Carrying cost components	Ranges per year, %
1. Interest on investment...............	5–6
2. Space charge.......................	¼
3. Handling charges...................	1–3
4. Supplies...........................	¼
5. Insurance..........................	¼
6. Taxes..............................	¼–½
7. Obsolescence......................	5–10
8. Depreciation.......................	5
9. Deterioration.......................	3
10. Use of money elsewhere.............	4½–8

experts showing both the factors involved in carrying charges and their possible magnitude.

Now try to figure your own carrying charges in one of the four following ways:

(a) ACCURATE METHOD: Have your accounting department supply data to fit the categories detailed in the table.

(b) PROFIT METHOD: If your firm is in a high-profit, fast-moving industry such as electronics, chemicals, or services, pick 36 percent (or 3 percent per month). This figure is OK too if your inventory is subject to high handling charges, obsolescence, or deterioration.

(c) INDUSTRY METHOD: If you are in metalworking, textiles, or general manufacturing, where profits are more modest and inventories relatively spoilproof, pick 24 percent (or 2 percent per month).

(d) TRIAL METHOD: If in doubt, pick 30 percent (or 2½ percent per month)—then refine this figure as you gain experience under EOV.

4. Figure your EOV. The above information, combined with monthly usage data, which can be obtained from either the purchasing or the production department, is all that is needed for substitution in the EOV formula:

$$\text{EOV} = \sqrt{\frac{2RC}{A}}$$

Since monthly usage generally varies item by item, mathematicians, by juggling this basic formula, have been able to simplify the calculation by putting all the variables except usage into a convenient table, of which Table 8-2 is a sample.

Its use can best be illustrated by following through on the example noted above where carrying charges were 2 percent per month, order cost was $4, and monthly usage was estimated at $10,000. The appropriate factor in Table 8-2 is 20.00 (where the $4 order cost intersects the 2 percent carrying cost).

TABLE 8-2 *Effect of Order and Carrying Costs on EOV*

Marginal cost per order or release	Carrying cost, % per month		
	1	1½	2
$1	14.14	11.55	10.00
2	20.00	16.35	14.15
3	24.50	20.00	17.35
4	28.30	23.10	20.00
5	31.60	25.80	22.40
6	34.70	28.30	24.50
7	37.50	30.60	26.40
8	40.00	32.70	28.30
9	42.50	34.60	30.00
10	44.75	36.50	31.60

Next take this factor and multiply it by the square root of the given monthly usage. The square root of $10,000 is $100. So 20.00 is multiplied by $100, giving the resulting EOV of $2,000. This is exactly the same answer calculated by direct substitution in the formula (see page 152).

One complicating factor arises where the EOV is relatively small, say, the equivalent of a one-month supply, but the lead time (time of order to time of delivery) is longer (say, three months). Under this kind of condition, it would be necessary to have more than one order open at any given time. But, note that this would in no way affect the basic EOV value.

By still further manipulation of the basic formula it is possible to figure how much of a penalty is incurred when you order uneconomically. Specifically:

$$\text{Penalty} = \frac{RC}{Q} + \frac{AQ}{2} - \sqrt{2RAC}$$

where Q is the dollar order being considered.

Assume in the above example that instead of ordering $2,000 worth of steel, we order $100,000 worth. Then:

$$\text{Penalty} = \frac{(4)(10,000)}{100,000} + \frac{0.02(100,000)}{2} - \sqrt{2(4)(0.02)(10,000)}$$

$$= 0.4 \qquad\qquad + 1,000 \qquad\qquad - \sqrt{1,600}$$

$$= 0.4 \qquad\qquad + 1,000 \qquad\qquad - 40 = \$960.40$$

In other words, had you ordered $100,000 worth at a clip, you would have been wasting approximately $960.40 per month.

These penalty computations are particularly useful when considering higher-than-EOV quantities because of quantity discount offers or the possibility of a volume freight savings. The rule of thumb: If the savings is greater than the penalty for buying more, you're ahead of the game. If not, stick to your original EOV order size.

Capital Spending

Probably one of the most important—and also most difficult—problems a company faces is making a decision on capital investment. Usually two questions are involved: (1) Is the outlay necessary to meet modernization and/or expansion needs? (2) If so, where, when, and how should the funds be spent?

Answers to these questions revolve around earning potential. Obviously, the decision that offers the best chance of maximizing profits is the one to pursue. Less obvious, however, are the procedures for calculating the profit potential. The simple methods leave much to be desired; the more complex ones are often time-consuming and based on much guesswork.

Generally speaking, the following are the three basic dollars-and-cents yardsticks for measuring the worth of an investment proposal.

1. **Payback period.** Stripped of its niceties, this concept is concerned solely with the question, How long will it take to get invested money back? It's generally quite popular for three reasons: It's measurable, it's easy to compute, and it's easy to explain. Thus if it costs $10,000 to buy a machine and the additional income after taxes is $4,000 a year, the payback is $10,000/4,000, or 2½ years.

Another advantage is that it gives a handy estimate of cash flow back into corporate tills. In these days of high working-capital needs, a proposal that guarantees a quick return of badly needed funds will usually be highly regarded.

Another plus factor: The length of the payback period is a valuable gauge of risk. A short payback period means that invested money will be returned in the near future, when revenue predictions have the highest degree of reliability and validity. On the other hand, a long payback period leaves a firm open to the risks of the fuzzy long-term future.

But one shortcoming is that this measure ignores the years after the payback period. Put another way, it stresses liquidity rather than profitability. Also, it penalizes innovations, where the long-run profits may be high but the payoff period is slow.

2. Rate of return—accounting method. This is a simple method designed to supplement the payback period by measuring overall profitability. It expresses average annual net income (after taxes and depreciation) as a percent of capital outlay. The key word here is "average," since income over the life of the machine is considered without regard to time.

The trouble is that while this approach has the advantage of simplicity, it ignores the fact that (a) a dollar earned today is worth more than a dollar earned 10 years from now and (b) income may not be evenly distributed over the life of the equipment. These latter considerations can make a big difference in figuring the true return on investment.

Take the point that a dollar earned from the machine today is worth more than a dollar earned 10 years from now. This becomes clear when you realize that money earns interest, and thus the promise of $100 today is worth considerably more than the promise of that same $100 around 10 years from now. Specifically, today's promise is worth almost twice as much under current interest rates (see Figure 8-3).

With the above in mind, it becomes obvious that two projects with the same total and average earnings can have totally different rates of return. The one where the bulk of the earnings occurs in the early years is preferable because of a higher rate of return, less risk, and earlier payback.

3. Rate of return—discounted cash flow. This method is specifically designed to eliminate the time difficulty, since it takes into account when the earnings are received. While somewhat more complicated than the other methods, it's by far the most accurate approach toward measuring investment profitability. As such, it has found widespread acceptance among major firms over the past decade.

The true or realistic rate of return on a specific proposal (determined by this method) can be readily compared with those of alternative proposals. And other things being equal, the equipment with the highest rate of return will get the nod.

The methodology is based on setting the revenues or cash flows for all future years against the cost of acquiring the new equipment—so earnings further out into the future are given less weight than those nearby. Then by referral to a proper compound interest rate table, it's possible to determine the rate of return implied by these cash flows.

EXAMPLE: Suppose that your firm is considering the investment of $10,000 in a new machine tool. Assume that based on current knowledge this will mean additional net cash inflows over the next 20 years of about $2,000 a year.

PROBLEM: What is the firm getting in the way of return from this investment?

A look at the proper compound interest rate table would enable you to compute exactly what interest rate or rate of return these future cash flows (if properly discounted) would imply. In the above example it would be about a 19 percent rate of return. The proposal is therefore attractive even though the payback period is five years.

The trouble with this formula in practice is that it requires substantial calculations to obtain each year's net earnings. It involves figuring out every year's operating improvement, cost improvement, cost savings, tax depreciations, and income taxes—plus a lot of work with interest tables.

But there's a way out of the box. Special formulas, assuming "standard" cash flow patterns over the years, are available—formulas which simplify the arithmetic and the estimating. The MAPI (Machinery &

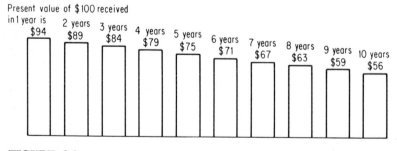

FIGURE 8-3

Allied Products Institute) approach discussed below is probably the most popular shortcut technique. Here's an example of how it might work: Your mill is considering the purchase of a $10,000 tool. This price includes shipping, installation, and sales taxes. Your old tool is scrap and worth nothing. Calculations show that next year you will make an additional $5,000 (through lower costs, a better product, etc.). Corporate income taxes are 50 percent. The machine will have a 10-year life with a 10 percent salvage value. Depreciation will be on a declining-balance method. What return will this new proposal yield?

STEP 1: Figure after-tax return on the first year's operations by taking 50 percent of additional revenue of $5,000. This gives an after-tax return of $2,500.

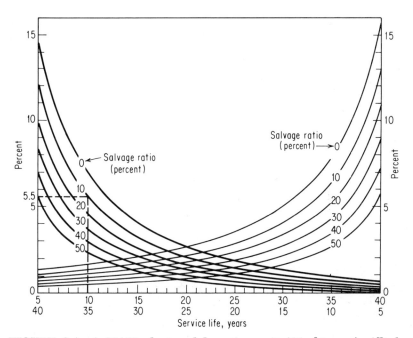

FIGURE 8-4 *A MAPI chart and how to use it* (*Machinery & Allied Products Institute*).

Instructions: STEP 1: Use heavy curves for sum-of-digits or double-rate declining-balance depreciation and light curves for straight-line depreciation. STEP 2: Locate service life (in years) on the horizontal axis, reading from left to right for heavy curves and from right to left for light curves. STEP 3: Ascend the vertical line to the point representing salvage ratio (estimate the location when the rate falls between the curves). STEP 4: Read the point opposite on the vertical scale. This is the MAPI chart allowance.

STEP 2: Refer to the MAPI chart in Figure 8-4. Note that the second heavy line from the top corresponds to declining-balance depreciation and 10 percent salvage value. Note that this yields approximately a 5½ percent MAPI chart allowance for a machine likely to last 10 years.

STEP 3: Apply this 5½ percent allowance to the installed cost of the project of $10,000. The result yields a dollar MAPI chart allowance of $550.

STEP 4: Subtract the dollar MAPI chart allowance ($550) from the after-tax return of $2,500. The result ($1,950) is the amount available for return on investment.

STEP 5: Divide the amount available for return on investment ($1,950) by the net investment required ($10,000). The result of 19½ percent is the rate of return, or profitability.

These calculations indicate that the new tool would yield a 19½ percent return on investment. This seems quite high—and certainly it would be better to buy the tool than to keep the money in a bank or invest it in high-grade bonds.

But before you make any final decision, compare the profitability with earnings available from alternative proposals that might possibly yield an even higher return. Also, the payback period (another investment criterion) should be checked to see whether it's within the limits set by your company policy.

Presumably, the discounted cash flow method or one of its relatively simple approximations such as the MAPI formula should also give helpful hints on the best time to invest. Thus the most propitious time would be when the true rate of return was highest.

Aside from decisions on individual projects, a few broad economic generalizations on the problem of investment timing may be in order. Very seldom will it pay to embark on a new expansion program when a business boom is approaching a peak, for in most cases the new plants will be coming on stream during a period of falling demand—hardly the best time for profit optimization.

American industry's record on this score leaves quite a bit to be desired. Thus a recent MAPI study reveals that during three recent cycles, business waited too long and invested too heavily in new equipment. As a result, about 55 percent of all factory construction during these cycles was completed after the crest of the boom.

Timing investment near the crest of a business cycle has another disadvantage: There are high fixed costs built into the investment, for boom times are traditionally times of high interest rates, high capital goods prices, soaring construction costs, etc. On the other hand, investing toward the end of a trough is usually less costly and starts yielding profits during a period of rising demand.

The danger of obsolescence is, however, a problem if plants are built a bit ahead of time. Clearly, then, the successful management is one that gauges the odds of technological obsolescence against the advantages of expanding during slack periods, when lower fixed costs can be built into the capital structure.

Input-Output Analysis

This is a relatively new marketing tool based on government tables which define, trace, and chart the myriad of interlocking buy/sell relationships existing between 370 broad-based industry groups. In short, these tables are designed to show what effect any industry's demand will have on the sales of every other industry.

While developed by longhair economists, these input-output (I-O) tables are fast becoming invaluable aids to industry forecasters and planners. One of the chief I-O applications is pinpointing markets for the future. Another use is profiling customers' market patterns to determine a company's market penetration.

At one large chemical company I-O analysis was one of the techniques used to develop the division's 10-year marketing plan. Using updated coefficients from the government's tables plus projections of consumer demands and the GNP, company planners worked back to determine demand for their base chemicals.

Buying strategy, too, is enhanced by use of these I-O tables. Thus purchasing officials can use the figures to (1) estimate future demand shifts for key materials, (2) check up on individual performance by comparing their purchase costs with those of the industry as a whole, and (3) measure purchasing leverage by comparing their own and their industry's consumption of a particular material with total consumption of the same material.

A recent General Motors strike can perhaps best illustrate still another use of I-O—that of estimating sudden shifts in demand. When workers went out on strike during the fall of 1970, it was widely recognized that this would have a severe impact on supplier industries. The big

TABLE 8-3   *How a 1970 GM Strike Hit Supplier Industries*

(Sales loss per day in millions of dollars)

1. Primary iron and steel manufacturing.................... 7.8
2. Stampings, screw machine products, and bolts............ 3.1
3. Other fabricated metal products........................ 2.5
4. Wholesale and retail trade............................. 2.1
5. Rubber and miscellaneous plastic products............... 2.0
6. Transportation and warehousing........................ 1.4
7. Primary nonferrous metal manufacturing................. 1.2
8. Metalworking machinery and equipment................. 1.2
9. Business services...................................... 1.1
10. Miscellaneous electrical machinery and supplies........... 0.9
11. Machine shop products................................. 0.9
12. Miscellaneous fabricated textile products................. 0.9
13. Auto repair and services............................... 0.9
14. Glass and glass products............................... 0.8
15. General industrial machinery and equipment.............. 0.7
16. Scientific and controlling instruments................... 0.6
17. Service industry machinery............................. 0.5
18. Paints and allied products.............................. 0.4
19. Electric lighting and wiring equipment.................. 0.4
20. Electric, gas, water, and sanitary services................ 0.4
21. Engines and turbines.................................. 0.4
22. Miscellaneous textile goods and floor coverings............ 0.4
23. Finance and insurance................................. 0.3
24. Business, travel, entertainment, and gifts................. 0.3
25. Radio, TV, and communication equipment............... 0.3
26. Gross imports of goods and services..................... 0.2
27. Stone and clay products............................... 0.2
28. Electronic components and accessories................... 0.2
29. Maintenance and repair construction.................... 0.2
30. Petroleum refined and related products.................. 0.2
31. Communication, except radio and TV................... 0.2
32. Real estate and rental................................. 0.2
33. Electrical industrial equipment and apparatus............. 0.1
34. Scrap, used and secondhand goods....................... 0.1
35. Chemicals and selected chemical products............... 0.1
36. Federal government enterprises......................... 0.1
37. Construction, mining, and oil field equipment............. 0.1
38. Lumber and wood products, excl. containers.............. 0.1
39. Heating, plumbing, and structural metal products......... 0.1
40. Household appliances.................................. 0.1

SOURCE: Commerce Dept., input-output tables.

question was, Just how big an impact? The answer—supplied by I-O analysis—is detailed in Table 8-3.

As might be expected, the steel industry stood to lose the most. But as the table indicates, a lot of other industries were seriously affected by the work stoppage.

It should be noted in passing that several simplifying assumptions were made to arrive at the results shown in this table. But this in no way detracts from I-O's big advantage in this case—its ability to come up with "ball-park" estimates in a matter of minutes. Surprisingly enough, the total (all-industry) impact suggested by this kind of I-O analysis came quite close to General Motors' own estimate.

While I-O tables are relatively simple and easy to read, they represent the end result of years of research and calculation—for the new I-O approach does a lot more than just estimate the direct effect on every industry of, say, a given change in the demand for passenger cars. In fact, this direct impact (how much steel, nonferrous metal, chemicals, etc., will be immediately ordered by the auto makers) is just the first step in a long line of repercussions.

Thus steel sales, derived from a boost in auto demand, in turn set off a secondary wave of sales; in order to meet increased steel sales to auto makers, the steel mills have to step up their own buying of chemicals (such as sulfuric acid), iron ore, limestone, coal, coke, and perhaps even some new equipment.

This "multiplier" process then continues in a tertiary buying wave as producers of chemicals, iron ore, limestone, etc.—influenced by greater sales in the steel industry—have to step up their own purchasing activity to meet increased steel activity.

This multiple-wave buying effect of an original boost in any industry's demand is traced for each of the 370 specific industries which make up the economy. As a matter of fact, the I-O analysis sums up virtually hundreds of thousands of these demand repercussions, thus allowing the forecaster to trace the complicated and highly intricate chain reaction through the entire industry structure.

Generally speaking, these chain reactions or secondary effects are quite important, though they tend to vary from one industry to another. In the steel industry, for example, a $1 increase in automobile demand results in direct orders to the steel industry of about 8½ cents. But when all the multiplier effects are added in, the final increase in steel demand stemming from the initial auto increase is nearly $2\frac{1}{2}$ times that amount—specifically, over 20 cents.

One doesn't have to be a mathematical whiz to work with the new tables. Says one government official who helped mastermind the entire project: "I could probably sit down with someone who has never seen input-output, and within an hour or so he would know how to use the approach." The heart of the technique lies in the understanding of three basic sets of tables, shown in Figure 8-5.

One criticism leveled at the approach is that there are hundreds of gaps in the I-O grid—particularly on product detail. While the sales volumes and sales forecasts of entire industries are valuable, a marketing man is most interested in product lines. The Department of Commerce grid gives figures for only broad industry categories.

1. The specific destinations of every individual industry's sales. Table (a) gives the percentage of a given industry's sales that go to every other industry or to final demand (direct consumption). Thus, in the simplified Table (a) below, some 10.3 percent of steel-industry shipments go directly to automakers.

2. How supplier industries' sales vary directly with changes in user-industry demand. Assuming a $1 change in the demand of a given industry, Table (b) gives the exact number of cents change in the sales of each industry directly supplying the given industry. Thus, in the simplified Table (b) below, a $1 increase in the auto output means a direct 8.5 cent increase in steel sales.

3. How supplier industries' sales vary overall (directly and indirectly) with changes in user industry demand. In addition to the direct effect on supplier sales, Table (c) also takes into account repercussions stemmimg from the fact that affected supplier industries, in meeting changing demand patterns, have to make changed demands on their own suppliers as well. Thus, a $1 increase in the final deliveries of autos means a total (direct and indirect) 20.3 cent boost in steel sales

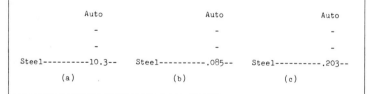

	Auto	Auto	Auto
	-	-	-
	-	-	-
	Steel----------10.3--	Steel-----------.085--	Steel-----------.203--
	(a)	(b)	(c)

FIGURE 8-5 *Input-output analysis tables.*

For example, it shows that the electrical industrial equipment and apparatus industry sold so many millions of dollars worth of its products to the new construction industry. But sales by kinds of electrical equipment aren't reported, and neither are purchases by types of new construction. (It should be noted that some unpublished details are available, however.)

To counter this, several commercial organizations are developing updated, more detailed I-O grids. Some are offering I-O-based marketing services for a fee. Their reliability, of course, depends on the expertise of the individuals who make the required judgments. And it takes an army of individual experts. No one person—or handful of people—knows enough about business and industry to do it alone.

Another problem: You cannot assume that the industries on the I-O grid are identical to those defined by the Standard Industrial Classification (SIC) system used in business censuses. In many cases, customary SIC groupings were dismembered and realigned for I-O purposes. So it is possible that a marketer who has been accustomed to thinking he is in industry A will find that his kind of product has been shifted to industry B in the I-O figures. This could mean that any relationships he's been able to trace between his company's sales and those of his SIC industry are not compatible with I-O data. To make matters worse, this could apply to his customer industries as well.

Still another criticism involves the time lag. For example, coefficients for 1963 weren't released until 1969. However, past experience indicates that these coefficients tend to remain constant. Also, with new computer techniques the government hopes to reduce the lag to two or three years.

NINE

The Role of the Computer

Most of the sophisticated statistical techniques and approaches now in common use have been known for decades. Basic probability theory, for example, was developed way back in the 1920s—nearly 50 years ago. Yet it has been only in the last few years that statistics has come into its own as a powerful tool—one capable of taking so much of the guesswork out of the business equation.

Why the delay? The answer is simple enough. Up until a few years ago business just did not have the physical means of implementing its theoretical knowledge. But with the advent of the computer, this all changed. Something that previously may have represented months of drudgery (if indeed it could be accomplished at all) now requires little more than a few minutes of time on an electronic data processing machine.

But it would be a mistake to say that the computer's value lies solely in its capacity for the application and analysis of sophisticated statistical techniques. To many firms an equally important dividend has been its ability to generate enormous amounts of useful raw statistical information—on prices, profits, production, inventories, etc. Indeed, a surprising amount of computer time is still devoted to such "pick and shovel" tasks—for without good underlying data, meaningful analysis is impossible.

Another big computer plus: Speedy data processing machines make it a lot easier to touch all bases—or at least many more than might otherwise be touched. Suppose, for example, that it is known that 50 or so variables may have an effect on steel sales. To do an adequate job of predicting steel sales then would imply consideration of a bewildering array of different equations. Without computer aid this would be a nearly impossible task, involving perhaps several man-months of work, and even then not all the pertinent questions might be answered.

A simple exercise in statistical probability can perhaps best illustrate the problem. Suppose that the analyst is toying with four possible variables, any one or a number of which can help explain fluctuations in sales volume. This seemingly simple problem presents the analyst with 16 possible alternatives: one way of considering all four variables; four different ways of considering any three of the four variables; six different ways of considering any two of the four variables; and finally, one way of rejecting all four possible variables.

The number of alternatives becomes astronomical as the number of possible sales-predicting variables grows:

No. of possible variables	No. of alternatives
4	16
10	1024
20	1,048,576
100	1.26765×10^{30}

It goes without saying that if the possible form of the relationship (linear, etc.) is viewed as another question mark, the number of alternatives becomes even more fantastic.

The point to be made is this: Most good forecasters find that there are many computations they would like to make, and yet for reasons of staff size, money, and time, they cannot make them. Bring in a computer, however, and the impossible becomes the possible.

Detail is another big plus. Computers can split up data into any number of useful breakdowns. Production control, for example, might want a breakdown by individual product, market research by industry group, finance by overall value and volume, sales by territory, and so on. Once the basic information is available on tape, it can be processed in any way and with very little extra cost or time.

Finally, there is the question of timeliness. Once a program has been set up, basic data or new forecasts can be sent to top management just hours after new, updated factors become available. Thus, if sales of cement depend on housing starts, then the sales projection can be completed almost immediately after, say, the government or some private group comes out with a new estimate of housing starts. To put it another way, computers assure the use of the most up-to-date information possible.

It should, however, be emphasized that this dependence on the computer in no way diminishes the status of the analyst. Quite the contrary. He becomes increasingly important—for the advanced techniques usually computerized are powerful tools requiring an enormous amount of sophistication in regard to both the techniques themselves and the data being analyzed. In short, it takes a lot more savvy to interpret the results of a computer-based analysis than to interpret those of a simple, hand-calculated one.

Indeed, if anything has stood out in recent computer usage, it has been the need for more and better-trained personnel. Both for deciding on input and for evaluating output, there is not now, never has been, and probably never will be any substitute for sound judgment.

Commenting on this point, an auto executive observed: "The computer is an aid to thinking—not a substitute for thinking. It extends brainpower—just as mechanical tools extend muscle power. Technical devices, whether they be machines or mathematical models, are only a link in the chain of individual and social action. Individuals must set the requirements and ask the questions, and only individuals can decide what to do with the products and the answers."

This official adds, however, that the statistical aid provided by the computer can be extremely valuable. He notes, for example, that his company has already successfully applied the systems-analysis approach to the development of more accurate short-range forecasts of customer wants in automobiles. These forecasts, he says, have helped the company determine its production mix of models and accessories and are serving as guides for dealers in tailoring their inventories to customer needs.

Perhaps the biggest plus of all in turning to the computer is the tremendous improvement in manpower utilization. Before computers came on the scene, most statistical analysts were deeply involved in computation. Now the picture has changed. The time-consuming, tedious aspects are fast disappearing. These same people are now free to think, which is essentially what they are being paid to do. They need no longer be what one sarcastic corporate president described as "a bunch of highly overpaid statistical clerks."

But again a word of caution. Just because computers do all the dirty work, there is a temptation to ignore one of the basic precepts of statistical analysis, namely, checking on the quality of the data being analyzed. The machine can do all the work, but it cannot pass judgment on the quality of the original input. What has been said before holds doubly true for computers: The value of a forecast, no matter how sophisticated, is only as good as the data that go into it.

Today's Number One Growth Area

Figure 9-1 should quickly dispel any doubts about the fantastic growth of computers in recent years. Note that overall outlays for this sophisticated equipment have been growing at twice the annual rate of pro-

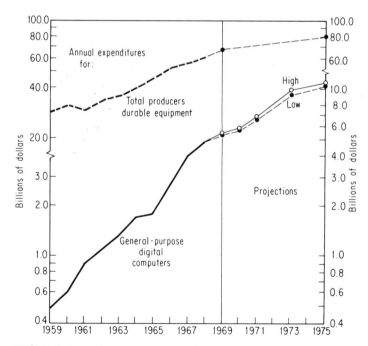

FIGURE 9-1 *Computers—a major growth industry* (*National Industrial Conference Board*).

ducers' durable equipment—a category which, in turn, has been moving up at a much faster pace than the general economy. Specifically, between 1965 and 1968 factory shipments of electronic data processing machinery and associated equipment more than doubled in value—with almost as fast a growth rate projected out into 1975.

In short, the computer business is the world's fastest-growing major industry, no matter what yardstick you may choose to use. For example, in 1956 only 570 computers were installed in the United States, with a total market value of $340 million. At the close of 1968 worldwide installation of computers totaled some 70,000 machines, with a purchase value of well over $20 billion.

Moreover, of the 1968 worldwide total, some 56,000 were installed in the United States. Approximately 14,300 of these were installed in 1968 alone, according to the Diebold Group, a management consulting organization specializing in automation.

Another major growth area is the development of peripheral equipment. According to one estimate, by 1975 the cost of the computer itself will be 30 to 40 percent of the entire hardware cost of the system, with the peripherals accounting for the majority of the cost.

IBM, the leader in the field, typifies industry optimism. IBM guesses (and it sounds plausible) that in a decade, computers will be the world's single largest business. Thus, even the smallest companies will soon own (or share) one of these electronic marvels, giving every business analyst in the nation access to more or less complete electronic data processing (EDP).

Moreover, IBM feels that the industry is still in its formative stage. The firm notes, for example, a new trend toward the use of computers in information networks that link many people at widely separated locations. This is made possible by the use of faster, more powerful computers and by the development of input-output devices that allow communication with the computer over any distance. It is generally agreed in the industry that this method of using computers will grow substantially over the next few years.

The mushrooming of companies in the field provides still further evidence of the computer's growth potential. Specifically, in the 1960s the number of computer manufacturers more than doubled, and by 1969 there were over seventy manufacturers of systems—some of them very large and well known—actively competing in the computer market. By the latter year there were another 4,000 outfits that offered peripheral equipment, programming support, and other services to computer users. Many of these firms have become very successful—some in just a few years.

An exhaustive 1969 study by the Research Institute of America further supports the growth thesis. The Institute found that at that time:

▪ Some 63 percent of its members—both large and small—were using computers.

▪ An even more impressive 82 percent had at least one executive who was familiar with EDP—and thus had the ability to use it effectively.

▪ The overwhelming majority of the companies with more than 200 employees or $7 million annually in sales were using computers.

And it's easy to see why. It is estimated that about 20 percent of the total output of services in our complex economy is now devoted to reports of one kind or another. This provides a ready-made market for computers, which can improve control over inventories, billings, market analysis, payrolls, taxes, and a host of other record-keeping administrative functions. One source, for example, estimated in 1969 that about half of the computers then in operation in the United States were devoted to the control and reduction of administrative paper work and cost.

Technological progress has also contributed to growth—with more efficient, comprehensive machines adding to the market potential virtually every single year. Indeed, in less time than humans can complete one generation, the computer has gone through three and is now in the process of starting a fourth.

Actually, the analogy between computers and humans isn't too bad. Each computer generation—like the human variety—inherits something from its forerunner. Each incorporates improvements and new sophistication (which one hopes is true of humans, too). And each overlaps the other. Here is a thumbnail description of what went on before, where we are today, and what is to come.

Computer generation no. 1.   This was the vacuum-tube era. The computer's insides consisted of vacuum tubes or relays 2 to 5 inches long. By later standards, these machines were limited in memory capacity, and their processing speed was comparatively slow.

Computer generation no. 2.   This marked the beginning of the transistor era. Instead of vacuum tubes, the computer was built around transistors, ½ to ¾ inch long. The upshot of this improvement was a considerable increase in both speed and memory capacity.

Computer generation no. 3. The characteristic of this era was microelectronic circuitry. By replacing either vacuum tubes or transistors with ultracompact minimarvels, computer makers were able to step up speed and memory capacity once more—and, of course, to keep the physical dimensions of the machines from getting out of hand.

Computer generation no. 4. Introduced in 1969 and 1970, this generation once more shows advances in speed, memory, and space/performance ratios. And it also has greater flexibility in programming techniques.

Equally important over the past few years has been the development of the minicomputer—sometimes called the desk-top computer or "mighty mouse." This has opened up the possibility of EDP to medium-sized and even small firms. Among the advantages for these smaller firms, the following four stand out:

1. Price: These babies generally go for under $10,000—well below the cost of their bigger brothers and sister. This makes the minis attractive to the great majority of corporations, which can't afford or justify a bigger model.

2. Capability: For many purposes it is surprisingly good—a small computer can do many of the things the big ones do.

3. Versatility: You can hook these minis to a large central computer operation and thus establish an excellent feed point.

4. Convenience: If you want a personalized service not readily available from a computer operation located outside your own company (such as a data center or time sharing operation), the mini might be the answer.

In all, the small computer should be taken seriously. It's not a toy.

On the other hand, you have to think of the minicomputer versus the big computer in the same way you would think about a small, simple camera versus the elaborate reflex models. Note these minicomputer limitations:

▪ The little ones tend to have little memories. This may necessitate writing programs in direct computer language rather than using program languages such as Cobol or RPG (see page 184). If that's the case, you have to get a really topnotch programmer to set up your system. Direct-language programming is not a job for amateurs.

▪ The minis are somewhat slower in processing.

- The availability of software is limited.
- Minicomputer manufacturers generally sell the basic unit only and often do not provide the software and services marketed by the makers of the big machines. In short, this means you are more on your own.
- If you try to upgrade the capabilities of a mini to those of a maxi, you're apt to acquire so much peripheral gear that you will get into the price bracket of the maxis without really achieving the same benefits.

Before leaving the subject of computer growth, a few further words on computer-based services may also be in order. This aspect of computers embraces such terms as "software," "computing utilities," "data banks," "multiaccess systems," and "time sharing." It's a business that grossed $2 billion in 1968, and it's growing by 25 percent a year now—twice as fast as computer equipment sales. In fact, analysts think it will get bigger in the next 5 to 10 years than the computer makers did in their first 20.

Computers are the capital equipment, not the product, in this field. The raw material is the paper-work glut and the information explosion. The end products are dozens of electronic data processing services for handling chores as varied as payrolls, designs for new bridges, and instant credit checks.

Moreover, since the name of the game is know-how, almost everybody can and does play. Unlike major production industries—most of which grew from the nucleus of a single invention, product, or raw material—computer-based services sprout up wherever a few brainy people with EDP savvy gather around a computer.

The incentive for using these services is almost irresistible. Multiple-access computer systems already provide sophisticated EDP resources that only a few very large corporations could afford to set up on their own.

For small businesses, the greatest computers in the world are now as accessible as the nearest telephone plug. For real estate men, for example, there's a data bank with instant readings on available properties across the country; for merchants, there are several credit-checking services; for hospitals, there's General Electric's Medinet package of data processing services.

Generally speaking, most computer services fall into one of the following four categories.

Data processing service centers.   These real pioneers of the business still account for the bulk of service revenues. The Association of

Data Processing Services Organizations (ADAPSO), which includes both independents and service arms of computer makers, figures that billings will grow by 25 percent or more a year.

Computer software houses.   These range from companies that design whole command and control systems for the Defense Department to "body shops" that simply provide programmers—and are the second largest sellers of EDP services in terms of dollar billings. Fees here cover management consulting and computer time for running problems.

Banks.   Financial institutions billed another hefty amount for "automated services"—the banker's term for the use of his computers to run his customers' payrolls, accounts receivable, and other jobs.

Time-sharing and other remote service operations.   These involve computer-communications hookups—and they are getting a lot of attention because of their dramatic growth rate, though their total revenue still lags behind the others'. But gains are expected to take off—and one recent survey forecasts that time-sharing service sales will be more than $1.6 billion by 1975.

One thing for sure, the young service industry is reshaping the internal structure of many businesses, amplifying the effect computers have already had. Not only is it capable of taking over many services that companies now perform on their own computers, but it will also add many new ones, simply because development, programming, and operating costs can be shared by many subscribers.

Computer Uses

As pointed out above, the use of the computer is almost unlimited—embracing virtually every phase of business operations where information can be quantified. This section will attempt to point up some of the more popular areas where the EDP approach has been applied with spectacular success.

1. *Banking.*   Given the cost and time pressures inherent in the existing settlements systems, it is not surprising that banks quickly adopted EDP to speed up the check-clearing process and lower the per-item handling cost. Nevertheless, it has become apparent that the long-term solution is not further automation of handling, but rather a reduction in the flow, leading to what is popularly called the "checkless society." A more realistic goal is what may be termed the "fewer checks and less cash" society. In short, thanks to the computer the new financing system is likely to be a composite of cash, checks, and electronically produced information.

A key step in the evolution of such a society has been the recent introduction of multipurpose bank credit-card plans. These new media of exchange, which would not have been practicable without the computer, are capable of providing revolving lines of credit for consumers and a guarantee for the merchant.

These new plans in many ways resemble those under which the major nonbank interests issue cards. For a small fee, the card-issuing bank (or group of banks) establishes relationships with merchants who agree to honor the card for retail purchases. The merchants receive immediate credit for all deposited sales slips which originate via the credit card. And when the customer subsequently receives his statement from the bank, he may remit the amount outstanding in full or pay it over an extended period.

This computer-oriented approach benefits consumer and merchant alike. In addition to reducing the number of checks for settlement, the plan relieves the customer of the necessity and inconvenience of frequent check writing or of carrying large amounts of cash. In general, these programs resemble those under which credit lines are granted to business borrowers. Applicants deemed credit-worthy are granted a stated line of credit—all or a portion of which may be borrowed simply by writing a check.

There are even bigger pluses for the merchant. First, the plan reduces his accounting and record-keeping costs. Second, it cuts his credit losses and collection expenses. Finally, since it reduces the amount of credit a merchant has to extend to his customers, it tends to cut his own need for credit.

2. Government. Virtually all periodic government reports—including production, price, inventory, and GNP surveys—are now processed with the help of computers. The resulting reports are more timely and more accurate, and in many cases they include far more detail than was possible in the precomputer age. In some cases, the federal government know-how has spilled over into the private sector. Thus the Census Department now makes available to businessmen time-series decomposition and seasonal analysis (see Chapter 7), which have provided valuable help in short-term company forecasting and analysis.

3. *Industry data handling.* Computers have proved of tremendous help here, for anyone who does systematic analysis of statistics is quick to find that his basic data are always changing. First, he is constantly getting more data as time progresses; second, the people who prepare the data for his use appear to take almost fiendish delight in revising their original figures. Usually this occurs just after a forecast has been completed. More figures and revised figures make it necessary to recompute statistical relationships from time to time. There is therefore the problem of keeping a data file.

In many firms, forecasters have found that the easiest and quickest way to do this is to keep the file on tape or punched cards. Although major revisions are sometimes made in the statistical series which make it necessary, for example, to throw out a card deck for the series and punch a new one, experience indicates that for a long time series, a considerable part of the deck can be saved when such revisions are made.

The data in these card files are usually kept in the form in which they are reported. But in using the data in statistical work, the analyst frequently wants them in other forms. He may wish to use the data in logarithms, in differences, or in differences of logarithms. The computer, properly programmed, can usually accomplish such transformations in a matter of minutes.

Moreover, the speed with which change comes is leading more and more concerns to set up entire systems of information in marketing. After four years of working on a pilot model, General Electric got its marketing information system working in earnest at the corporate level early in 1967, for both its consumer and its industrial groups.

Essentially, GE has set up its own data bank, utilizing its own national sample, which it samples once a month on a continuing basis or whenever else the occasion demands. This sample is a primary source for the consumer inputs. A far-flung network of information from a variety of sources serves as the data for the industrial research.

The above type of data bank is often referred to as a "horizontal" one since it provides a wide area of detailed information for a single point in time. Equally popular these days is the so-called vertical bank, which usually covers a selective list of business indicators over an extended period of time. One advantage of this latter type is that it is uniquely designed for use with a time-sharing computer system.

First National City Bank's experience is a case in point. Using a time-sharing approach to storing economic data, the big New-York-based bank gave this appraisal in 1969:

> Currently, there are about 500 to 600 series in the data base, and FNCB's bill for pooled storage—that is, our proportionate share of storage charges for the whole data base and allied programs—came to less than $60 in the latest month billed. That works out to 12 to 14 cents per month per series for having prompt access to a historical statistical series, 1947 to date, which has been checked, updated, and entered into the computer. When you figure the cost of duplicating that from scratch, even at a clerk's wages, let alone those of a researcher or economist, you can see that you don't have to access many series before the data bank begins to pay off.

4. Market forecasting. This is another computer area with tremendous potential. To be sure, a lot remains to be done here. But already the payoff evidence is impressive. Thus a recent survey of more than one hundred top marketing vice-presidents showed increasing use of the computer for research and forecasting. Specifically, two out of every three of these executives said they were applying computers to make sales projections. Most said such applications were particularly intriguing because they permit greater detail and sophistication at both the input and the output ends of the spectrum.

Thus one big equipment producer, with the help of the computer, now makes long-range sales forecasts for (1) each industry served by its twenty-five or so company divisions and (2) its own sales to each of these industries—broken down by company division. The end result: Product managers have a pretty good idea of their markets, and corporate management has a gauge of its overall stake in each of its customer markets.

Before the computer came along, adds another market analyst, forecasts were made mainly for total product lines, for a year as a whole, and for the entire country. Corporate headquarters would then have to break this down on a subjective basis for individual products and sales territories. But more often than not, problems arose. While the overall forecast might be fairly accurate, the individual parts were usually way off. Thus one plant manager might find himself with a warehouse full of unsold goods, while another might be losing customers because he couldn't meet burgeoning demand.

Computers have now changed all this—by permitting individualized forecasts by product, by sales territory, and on a weekly or monthly basis. Ergo, a plant manager can now plan ahead with reasonable confidence that he'll come out on target, or perhaps only 1 or 2 percent off the projected level.

One of the nation's largest steel mills has carried the computer approach even further—developing two independent EDP sales forecasting programs. One stresses demand from individual markets, and the other stresses demand for individual product lines. In addition to providing forecast details that are mutually exclusive, the two programs run simultaneously and are used to cross-check each other in total.

5. *Input-output.* Computers also permit greater use of newer analytic tools such as I-O. Creation of an I-O table matrix (see Chapter 8) requires the spelling out of the complex interrelationships between all sectors of the economy. To do this on a desk calculator would be literally impossible. Some firms with computers have even built up internal I-O matrices which show, by year, buyer and seller relationships in the various company divisions.

6. *Linear programming.* This is another technique that would have been impossible using manual methods of calculation. As described in Chapter 7, this approach provides a reliable means of assessing the best way of tackling a problem by trying a number of possible alternatives and finding the best one. Another technique discussed in Chapter 7 which has widespread computer application is PERT (Program Evaluation and Review Technique)—a method for planning, scheduling, and controlling the network of activities in large or complex projects.

7. *Tax collections.* Computers are also widely used in income tax collections—with electronic equipment used to check each return against all reports of payments from all sources, including interest from banks, dividends from stocks, and wages and salaries. Returns are also checked against predetermined guidelines for medical, charitable, and other deductions. If the return doesn't meet specified standards, it is flagged for further checking by Internal Revenue Service personnel.

8. *Inventory management.* Statistical inventory management is also ideally suited for computer usage. To be sure, some maintain that the use of statistical formulas is not justified because the added costs and complexity are likely to offset the probable additional savings. However, with the low cost of computerized calculations, the expense of statistical inventory management is relatively small compared with potential savings.

Probably the most conclusive way to establish the magnitude of such savings is by inventory simulation based on actual past occurrences. That is, several past periods can be employed to establish a demand forecasting model for a representative sample of inventory items. Then this model (which should include the statistical inventory management techniques described in Chapter 8) can be applied to the actual time-varying demand for each item in new and untested past periods. A comparison of the new techniques with the previous best techniques under actual operating conditions will give an unbiased measure of the achievable savings. Typically, the new techniques reduce both stockouts and inventory levels for an item.

But computers can be utilized in inventory management even where no such sophisticated statistical approach is employed. Here's how one firm uses the computer in a basically mechanical way: The computer receives monthly forecasts from marketing. It next explodes bills of materials through the system, constantly updating material profiles on a piece-part demand basis. As parts needed for production come to their lead-time cycle, the computer issues a purchase requisition indicating time of order, quantity of order, and time of delivery.

The computer-generated purchase requisition card goes to the inventory planner, who either approves it as it was generated or changes some data.

The inventory planner then feeds the purchase requisition card back into the computer. The information on the card remains in the system as an open purchase order. Another card from the computer then goes to the purchasing department. A purchasing man negotiates price with the vendor and specifies how the order will be delivered. The clerical staff types the actual purchase order, and purchasing then feeds the actual price, method of delivery, vendor, and order number into the computer.

Between the time the order is placed and the time it is delivered, the computer issues daily commitment reports listing all goods that have been ordered but not delivered. A price-variance list is also produced. It indicates the difference between standard price and actual purchase price. This list is grouped according to buyers so that management can measure performance.

Three weeks before the order is due for delivery, it appears on the system's expedite report.

9. *Plant maintenance.* Computerized systems have been developed which can cut downtime, reduce overtime costs, monitor potential trouble spots, and allow better scheduling of work assignments. Not only do such systems produce schedules, vouchers, and reports, but they also automatically adjust to any changes in production equipment utilization. Thus if a second shift were to be added to a machine, a single keypunch would automatically double the frequency of future scheduled servicing.

Make no mistake about it—maintenance is now a major cost area. In some industries, for example, outlays range as high as 10 percent of sales, making the old catch-as-catch-can approach a luxury that few firms can afford in today's competitive markets.

The above listing of computer uses is in no way meant to be an all-inclusive one. Rather, it is presented as an indication of the myriad ways in which computers can be utilized in today's highly complex business operations. Indeed, there are literally hundreds of applications where computer usage has resulted in considerable dollars-and-cents payoff.

Finally, before leaving the subject of usage, it is important to distinguish between computer usage in production and distribution processes and that in the collection, processing, and dissemination of information. This book is concerned mainly with the latter—though the two functions may overlap.

Looking Into the Computer

So much has been written about the superhuman powers of a computer that it is often hard to separate fact from fiction. Perhaps the best approach is to start with the most basic definition of this new twentieth-century phenomenon. Specifically, a computer is a machine or group of machines linked together for the purpose of handling information.

This machine or group of machines generally consists of three basic units: (1) An input unit, where information that has been converted to machine-readable form (punched cards, perforated paper tape, magnetic tape, and so on) is fed into the machine; (2) a processing unit, where the information is processed; and (3) an output unit, where the processed results are presented for human inspection. This output may be printed on paper, displayed on a visual display unit, or even sent over a telephone line in voice form; or any combination of these may be used.

The information to be fed into the computer may be of many kinds, provided it can be coded into machine language. Thus names, addresses, weights, measures, mathematical formulas, etc., would all qualify for computerization.

The basic point to remember is that the great advantage of the computer lies not in its ability to perform complex operations but rather in its capacity for carrying out very simple tasks at incredible speed. Today's largest machines can execute millions of instructions in a second. The operations a computer can perform are, in most cases, limited to such simple functions as addition, subtraction, multiplication, division, and comparison of two quantities.

This ability to compare two quantities and to perform different actions depending on the results of the comparison is the key to the apparent decision-making power of the computer and is the main difference between computers and all previously used types of calculating devices.

This can probably be best illustrated by an example from the banking field. Consider, for example, the problem of whether to bounce a check or to accept it. The computer would attack the problem by first posting the check to the customer's account. Once this was achieved, it would compare the resulting balance with zero. If the result was zero or greater than zero, then all would be in order. On the other hand, if the result was less than zero, the computer would turn to a set of instructions that would reject the check and print a message stating that sufficient funds were not available.

But the computer, to perform any such tasks as accepting or rejecting a check, first has to be instructed. The approach is to set up what are known as *computer programs,* or sets of instructions on what the machine is to do. These are needed to establish a routine for the electronic circuits to follow.

Such programming is really the heart of computerization. Sometimes it takes weeks or even months to work up a particularly complicated program—as contrasted with the actual forecast based on the program, which might take only a few minutes.

How are such programs acquired? Many large firms have their own set of experts, and others call upon computer manufacturers, who will either provide some standard machine programs or help work out tailor-made ones with the user. Consultants are another good source for the development of meaningful programs.

In many cases, however, consultants are not needed, and the standard machine programs may prove sufficient. They are available now for such tedious work as multiple correlation, stepwise regression, curve fitting, making sales forecasting models, cyclical analysis, frequency distribution analysis, and seasonal analysis. Seasonal analysis is particularly valuable in evaluating time series where monthly or even weekly patterns have to be isolated before the underlying trend and cycle can be analyzed and extrapolated. These types of programs can be applied to big and small companies alike, to soft goods, to industrial products, and to consumer products.

Next consider the processing unit. This is the heart of a computer system and contains three main parts: the control unit, whose circuitry determines the sequence of operations and causes the computer to call for and execute program instructions; the arithmetic unit, where additions, subtractions, and so on are performed; and the memory or store, where the program, together with the information that is being processed at any given time, is held.

The program or set of instructions is placed for execution in the computer's internal store. The control unit signals for the first instruction in the store to be read out and acted upon. As soon as the operation is accomplished, a signal goes back to the control unit, which then signals for the next instruction. This procedure continues until the entire program has been completed.

This entire process is for the most part accomplished by use of the binary counting system. This is a number system based on the number 2 rather than the more common 10, used in the decimal system. In the binary system there is no actual symbol representing 2. Only 0 and 1 are used. For this reason the binary approach is particularly well suited for the computer since 0 and 1 are analogous to an electrical circuit with the current off (0) and the current on (1).

A simple illustration on binary counting should make this clearer. The symbol for zero in binary counting is the same as in decimal counting—0. The symbol for number 1 is also unchanged—1. It is only when we get to the symbol for number 2 that the notation changes. To create the equivalent of 2 in binary counting the symbol 1 is moved over one digit to the left. That's because under binary rules, any time the symbol 1 appears in the second column to the left, it stands for number 2. Thus the binary symbol for number 2 appears as 10—translated "2 plus 0." Similarly, number 3 in binary language is written 11—the equivalent of "2 plus 1."

Table 9-1 converts decimal counting into binary counting for all numbers from 1 to 25. This same approach, of course, could be used to designate numbers running into the billions and trillions.

TABLE 9-1

Binarys	Decimal
0	0
1	1
10	2
11	3
100	4
101	5
110	6
111	7
1000	8
1001	9
1010	10
1011	11
1100	12
1101	13
1110	14
1111	15
10000	16
10001	17
10010	18
10011	19
10100	20
10101	21
10110	22
10111	23
11000	24
11001	25

In terms of the electrical circuitry used in computers, the same type of approach would be achieved by turning the appropriate currents on and off. Thus, number 2 in binary counting would be signaled by having the left-hand circuit open but the right-hand circuit closed. Following the same reasoning, number 3 would require that both circuits be left open.

The use of on or off circuits, of course, needn't be limited to counting. Say, for example, that you had the problem of converting the word "dog" into terms of 0 and 1. Under this on-and-off circuitry approach you would break every letter up into 0-and-1 combinations (e.g., "a" is 010001, "b" is 010010, "c" is 010011, "d" is 010100, etc.).

This isn't always feasible commercially. So the computer people have invented internal conversion systems for computers which automatically do all the 0-and-1 work for you. However, to operate the conversion system, you first have to instruct the machine in a special "language," which—very loosely speaking—sounds like a combination of English, shorthand, and mathematics.

Note, too, that one conversion system won't translate everything from poetry to payrolls into 0-and-1 combinations. You need specific systems for specific chores. Hence, you also need specific languages to command the various systems.

Here's a rundown of some of the major languages—together with the jobs each one can do.

Cobol, *or "common business-oriented language."*   This one probably meets your requirements best. It was developed by a national committee of computer manufacturers and users, and it consists of easily learned English words and phrases (but don't bother to learn them—that's the computer people's job).

Cobol is intended for all common business applications—payrolls, cost accounting, purchasing data, etc. It fits most large computers.

Fortran, *which stands for "formula translator."*   As the name implies, this language is used for problem solving. It is written in mathematical and algebraic expressions. So if you want to do any forecasting, model building, or sophisticated calculating, Fortran may be your best bet. It, too, fits most large computers.

PL/1, *which means "program language/one."*   This is a sort of jack-of-all-trades language for a broad range of business and scientific work. But bear this in mind: Presently, PL/1 can be used only on IBM equipment (IBM developed it for its third-generation gear).

RPG, *or "report program generator."*   Simple English words and phrases are used in this one. Furthermore, you can get preprinted input instruction sheets which eliminate the need for step-by-step, detailed instructions. RPG is intended for smaller computers.

Coding

The development of computer languages goes hand in hand with the not inconsiderable problem of coding. For electronic data processing is virtually useless unless plant, vendors, customers, component parts, etc., can be translated into a code that both computers and working personnel can easily understand.

Suppose, for example, that you have access to a computer and you decide that it should keep track of a supplier named Bill Smith. Now, of course, inasmuch as the computer doesn't "understand" English, it is necessary to assign a number—say, 1000—to stand for the name Bill Smith.

But right here you can create two mammoth headaches unless you plan carefully in advance. One danger lies in not devising a proper numbering (or coding) system from the very start. And the other is the confusion that might arise in interfacing departments and divisions if they don't get the message that Bill Smith has been transformed into 1000.

Codes can be of many types. Some of the more common ones are vendor or customer codes, commodity codes, and plant codes. In many instances it is also possible to make some use of the government's Standard Industrial Classification code, or SIC code. It's particularly useful for compiling broad industry-oriented statistics on sales, inventories, prices, etc. In short, the code enables you to hang a common numerical label on the automotive end of business with General Motors, Chrysler, or Ford, for example. By the same token, that numerical label could also be used—broadly—as a commodity designation for autos.

Even more significant, the code is devised so that every time you add a digit to any basic two-digit designation, you refine the description. For example, the major code number for machinery, except electrical, is 35. Now suppose you add a third digit. Here's what you get: 351 is engines and turbines; 352 is farm machinery and equipment; 353 is construction, mining, and materials-handling machinery and equipment; and so on through 359, which is miscellaneous machinery, except electrical.

Next, take one of these three-digit figures—353—and add a fourth digit. Here's the refinement you produce: 3531 is construction machinery and equipment; 3532 is mining machinery and equipment, except oil field machinery and equipment; and so on through 3537, which is industrial trucks, tractors, trailers, and stackers.

And that's only the beginning. Breakdowns extend into six- and seven-digit codes. Moreover, there are provisions for 89 major commercial codes and several major government codes.

Sometimes an industry association can help in developing a tailor-made coding system. Fasteners are a case in point. The Industrial Fastener Institute recently—after more than six years of study and testing—came out with a product code consisting of a 31-digit identification number.

The code, according to the IFI, isolates eight specific fastener characteristics including product family, diameter, length, thread, class and fit, raw material, mechanical specification, and finish specifications—as well as assembly, packing, and packaging.

In releasing the code, the IFI stressed that it was compatible with electronic data processing and could be used effectively to simplify inventory control, production planning, material purchasing, or other information. Summing up, the trade association said: "Design, engineering, manufacturing, inventory, sales, purchasing, and accounting people—whether supplier or user—can now talk fasteners in a common language."

Embarking on a Computer Program

You can't swing into a successful computerized operation overnight. It takes a lot of blood, sweat, and tears. And what is appropriate for one firm may be unrealistic or actually misleading for another. Nevertheless, certain broad-gauge basic steps can usually start you on the way. Among other things, the following are important:

1. *Get a clear overview of what a computer is and does.* Otherwise, you will have serious communications problems with the specialists who run the machines—to say nothing of chaotic chores for your system. Some good outside sources for this kind of information would include courses given by computer manufacturers (though these are sometimes limited to customer company personnel) and visits to companies already operational in EDP programs.

Association materials and facilities provide another good learning source. Thus the American Management Association offers several courses—depending on where you are on the knowledge ladder—on the basic problems of computerization, including types of information needed, coding, computer capabilities, planning, etc. Sometimes more specialized types of associations can offer you detailed help in a special phase of computer activity. Thus the National Association of Purchasing Agents gives a special course on the use of EDP in procurement operations.

2. *Examine your present manual methods to see how efficient they are.* If the system you now use is too narrow in scope or too lackadaisical, forget about computers. You would merely foul them up. If you wanted, for example, to computerize purchasing operations, it would be important to ask these questions:

Can I analyze my present purchases by size of order?

Are inventory controls adequate in my establishment?

Do we keep proper vendor data on deliveries, etc.?

Do we have overall purchasing statistics on (*a*) expenditure of dollars by commodity, plant, vendor, and buyer; (*b*) cash discounts; and (*c*) foreign purchases?

3. If you have or can acquire adequate knowledge on the foregoing and if you feel the computer can help you in these areas, then start to think big. In the case of purchasing, for example, ask questions about the use of the computer for such chores as computer purchase orders (instead of requisitions) if the value of the items is insignificant, computer-generated requests for quotations, computer expediting and vendor monitoring, computer-compiled statistics, etc.

As for computer-compiled purchasing statistics, here are some guidelines you might possibly want to consider:

▪ Clearly define the type of information you think is really essential. It will probably include the following: number of orders, value of orders, value by commodities, number and value of orders by buyers, cash discounts, purchases outside the United States, and orders broken down by vendors and by plant location.

▪ Establish source documents. Some possibilities are a copy of the purchase order, the paper-tape by-product of your flexowriter output, and a copy of the invoice. Incidentally, be sure there's no hitch in the transmission of these data to the computer room.

▪ Establish frequency of reporting. Normally, semiannual or annual reports will suffice for vendors and commodities. Other times, a monthly frequency is desirable. Consider closing the purchasing month on the twentieth to avoid conflicts with departments closing on the thirtieth.

▪ Review closely with your systems and computer people. Good communications right from the start will pay off.

▪ Run a dual manual and dry-run computer system for three months. This will give you a check on the accuracy of the computer work, will pinpoint areas in which you need modification of the computer program, and will assure you that once the system is debugged, you can rely on computer figures. In all, this is a good way to avoid going off the deep end.

While the above are oriented toward a purchasing operation, they can also be used as a guide in setting up a computerized statistical intelligence for sales, production, or any other business operation.

Finally, in any decision about whether to jump on the computer bandwagon, there is always the question of how far to go, how much to invest in EDP. Some people try to relate this to the size of the company or to what competitors are doing. Unfortunately, there is no such simple rule of thumb. Outlays should be geared to the benefits derived in each individual case. And it should always be kept in mind that benefits accrue not only from the direct savings in data processing operations but also from the qualitative value of the information to be supplied by computerization.

This latter point will be discussed more fully in the following section.

Limitations and Problems

Like any business tool, computers present their share of headaches, dilemmas, and problems. Below is a list of some of the more common ones encountered by the typical industrial firm:

1. *Expecting the impossible.*   Many neophytes tend to attribute superhuman powers to these giant brains. Such computer mythology has in certain cases led to overexpectation—and ultimately disappointment. The key fact to remember is that the computer is not an intelligent human worker, but only a machine. Like any other machine, it does only what it is directed (or programmed) to do.

Disappointments can stem from many different factors. In some cases poor input is to blame. If the original input is bad, so is the output—no matter how sophisticated the processing. Another source of disappointment is overambitiousness on the part of the computer makers. Language-translation programs—despite glowing predictions by producers—ran into trouble a few years ago and even now are advancing at a much slower pace than originally envisioned.

But perhaps the keenest disappointments have stemmed from total dependence on quantified data for decision making. Too often, one tends to forget that computers, since they work only with quantitative data, tend to overlook equally important qualitative aspects of a problem. Thus—as noted earlier in the book—while the favorable "kill" ratio in the war in Vietnam, as analyzed by computers, suggested we were winning, the actual events proved that this was far from the truth. In short, the computer was incapable of evaluating all the subjective factors that go into such a complex undertaking as waging a war.

Another—but less tragic—example of this kind of statistical folly occurred a few years ago when the United States government was wavering about a decision on whether to buy or rent computers. Turning to the computer itself for the answer, they received a go-ahead to buy. Unfortunately, the computer, while it could quantify the dollars-and-cents cost savings involved, could not quantify the hard-to-predict obsolescence factor. The upshot: The government bought its computers, and shortly after that the industry came out with a new generation of computers that made its purchase woefully inadequate.

2. Poor utilization.   Coming back now to what the computer can do—there is often a tendency to put the emphasis on less important problem-solving areas. The net effect is that the user does not get top mileage out of his big computer investment. Specifically, the tendency is to stress cost displacement—but to ignore or give only minor weight to gains that increase management's capacity to control and plan. To be sure, reduction of clerical costs is important—and should not be downgraded. Nevertheless, this is not nearly as crucial or potentially valuable to management as the computer's contribution to policy-making decisions.

The problem is the usual one: It is easy to evaluate dollars-and-cents savings in clerical work, but not quite as simple to evaluate operational gains such as faster management reaction to a changing business trend. The latter is essentially an intangible savings. It is hard to put a dollars-and-cents value on any gain or advantage that might ensue. Other such intangible operational benefits of computer management include better customer servicing and improved forecasting techniques.

3. Investment evaluation.   The difficulty of measuring the operational advantage of computers in turn makes it harder to give a go or no-go signal on a particular computer investment. Specifically, the return on investment (ROI) criterion, so useful in buying production equipment, is not entirely satisfactory in this case. That's because ROI calculations are based solely on cost-displacement savings. They tend to ignore the equally important operational savings—and hence can often lead to the wrong investment decision.

There is a way out—albeit not an easy one. Rather than emphasize the cost angle, put the spotlight on the positive value of the information likely to be generated. Put another way, calculations should be based not so much on how much data generation costs as on its value to management. Admittedly, the latter information is a lot harder to come by and to quantify. But without it, any investment decision on whether to buy a computer is being made on highly questionable grounds.

4. The management-technician communications gap.   There is also a pressing need to bridge the chasm between the computer technologists and top management. Many technicians seem to feel that their top bosses are persistently ignorant of computer science and fail to set the objectives of the systems they have ordered their own people to build.

To correct this situation, data systems specialists hope to spread the computer gospel to the men who make decisions. This means grappling with such abstract questions as what information top executives want and need—and what executives and data managers should know about each other's jobs. It is also important not to overlook the requirements of auditors and accountants, a group of people who need different information from that required by operating executives.

Top-management spokesmen, meanwhile, are beginning to realize the extent of this communications gap. For the first time there's the recognition that technicians should not be setting management information goals. They rightfully point out that line officers should be the design architects.

Surveys also tend to point up the problem and its growing recognition by top management. One such recent survey was taken by the Diebold Group Inc.—a pioneering firm in the application of EDP techniques to business problems. Covering more than twenty-five hundred executives, the survey found that more often than not, it was the technicians rather than management who were setting computer goals—and that management was becoming increasingly disturbed about it.

The type of people entering the computer field adds to the overall communications problem. It should be remembered that the skilled computer specialist is today generally a young man who has grown up with this young profession. His average age is well under forty. Management has sometimes looked at this man as an upstart who is brashly overconfident in seeking to solve all corporate problems with his newfangled techniques. Some speak of this as a real generation-gap problem.

And, when the computer systems manager is placed on a low level on the company organization chart, he finds the usual manager's job of trying to sell ideas more difficult. In many cases his sales pitch must be convincing enough to sway as many as three levels of management, and this can reinforce his image of overaggressiveness.

All too often, the result is a formidable communications barrier between top management and the computer systems man. Apathy finally sets in on both sides, leading ultimately to high turnover rates for computer personnel and a low return on the company's investment in its system.

5. *The technicians' own communications gap.* EDP personnel have difficulty communicating not only with management but often with one another as well. The trouble stems from the tendency of each computer group to think in different terms. An official of the Home Loan Bank Board, lamenting his experience with the computer in this area, put it this way:

> It seems that those concerned all speak in different languages—not only with reference to the user but even right there in the computer complex.
>
> Analysts speak a jumble of fast-changing computerese when discussing what little they have come to understand of their projects. Programmers actually begin to think—and I have heard at least one habitually speak—in languages like Cobol.
>
> Input-output staffs, meantime, are thinking about the source data and the tables and graphs which they must review and approve. Operators tend to deal with the hardware—the dials, buttons, and routines of processing.
>
> Data librarians are involved with reel numbers, disk packs, tapes, retention dates, and data journals. Users, of course, find themselves unable to speak to or understand anyone in the complex, except perhaps the review clerks, who are really in no position to help them.

The real surprise is that this lack of internal communication between the various computer echelons hasn't led to even more chaos than it has.

6. *Manpower shortages.* Closely allied with the communications problem is the fact that there just aren't enough trained technicians around to meet the rapidly growing demand for EDP processing. This, in turn, has contributed to operating difficulties and costs that are much higher than necessary. Requirements for systems analysts and programmers, for example, have been growing at an annual rate of close to 25 percent. The supply is hard-put to keep pace. As a result, highly talented systems analysts and programmers have been difficult to find and even more difficult to keep.

As for holding onto them, one large company, which might otherwise have been proud of its $40-million investment in computer hardware, instead takes a "we don't talk about that" attitude because the turnover rate among its computer staff has been 80 percent a year.

Another company, which spends an average of $2.5 million a year on computer hardware and more than $3 million a year on systems unit salaries, underwent a 100 percent turnover in computer personnel following the resignation of the vice-president in charge of the operation.

And these are not isolated cases. A recent survey of 250 computer systems professionals (see Table 9-2) shows that by the time the typical computer systems man has been employed for ten years, he has changed companies more than three times and has changed to companies in different industries more than twice.

TABLE 9-2   *A Profile of the Computer Technician*

Age group	Average years of experience	Average earnings	Average number of company changes	Average number of industry changes
25–27	4.1	$13,427	2.1	1.9
28–29	5.9	$15,581	2.2	2.1
30–31	7.6	$17,012	2.4	2.1
32–33	9.7	$17,378	3.2	2.8
34–35	11.1	$19,995	3.5	2.7

SOURCE: Survey of 250 computer systems specialists, conducted by Edward Warren Org., New York.

Of course, shortages aren't the only factor behind this kind of turnover. Interviews with job-seeking computer men show that dissatisfaction with the role that companies expect them to play is the real reason behind the urge to move elsewhere. Many companies seem to consider the computer man simply another piece of hardware rather than someone who can join in managerial decision making. Change this attitude, and this fantastic rate of computer personnel turnover can be reduced. One company has found that since it reorganized its computer systems department and gave its staff a more challenging role in decision making, the unit's turnover rate has dropped from 70 to 18 percent.

Another manpower problem is that the growing complexity of fourth-generation computers will require that senior people fill a greater and greater proportion of these positions. Generally speaking, the increase in sophistication will require (1) greater emphasis on operations research and management science personnel, (2) more stress on systems analysts who can apply management decision concepts to company activities,

(3) more programmers who know how to use the sophisticated techniques of time sharing, teleprocessing, etc., and (4) the development of a new breed of professional EDP managers. The latter is crucial—for effective planning of the EDP function will undoubtedly be the biggest challenge in the years ahead.

Adding to the problem is the fact that—as in any rapidly growing area—few useful standards exist for the selection, classification, compensation, and training of computer personnel. Similarly, there are few yardsticks for measuring performance. The inevitable result: poor quality and soaring costs. The need for improvement here is again crucial—for computer personnel are fast becoming the major cost element in EDP systems.

TEN

Financial Statistics

More and more corporations are using their annual reports to peel the wraps from operating data they once considered taboo for competitors, analysts, and even stockholders. What's more, they are setting a trend that is expected to intensify over the next few years. Comprehensive reports, replete with charts and tables, are today providing businessmen and investors with a wealth of new information—data which can give considerable insight into sales and earnings performance.

This new trend can probably be attributed to a variety of causes—tougher Securities and Exchange Commission (SEC) regulations, more stringent stock exchange rules, and the realization on the part of corporations that only through full disclosure can they gain the investor's confidence needed in today's competitive battle for new capital.

Greater disclosure, in turn, has put greater emphasis on proper interpretation. There are now many more areas to monitor, and hence more places to go wrong. Also, like all statistics, financial figures are not always what they seem to be. Thus despite the tremendous reporting improvement of the past few years, there are still scattered instances of outright intent to hoodwink the investor (this is discussed below). More often than not, these involve errors of omission rather than commission—deliberate moves to leave some pertinent bit of information out of a report.

Conglomerates, in particular, often have been singled out for lack of operating information. Offended by what was described as one such firm's "expensive art catalogue," one acid-tongued critic termed the conglomerate's report an "ostrich-like" statement that revealed next to nothing about day-to-day operations.

But even with conglomerates, this secretive attitude is changing. The president of one such firm—one of the nation's largest—recently explained his company's change of heart this way: "It is better to let investors know in advance when things aren't going well in a certain area than to surprise them by suddenly selling off that operation." The unknown, concludes this official, is what really scares off investors.

This trend toward increased financial information is noted even in reports of overseas firms. Until recently, anything even remotely resembling full disclosure by European outfits was regarded as the equivalent of undressing in public. But, here too, this attitude is gradually changing (for the same reasons it is changing in the United States)—though it should be emphasized that overseas outfits still have a long way to go before they reach the level of reporting done by American firms.

In short—whether a firm is small or large, single-product or conglomerate, domestic or foreign—there has been an upsurge in the demand for better financial reporting, from the government, from stock exchanges, from financial analysts, and from purchasing people who must know about the production and financial capabilities of every vendor they deal with.

On the latter score, purchasing executives in recent years have used such financial intelligence for ferreting out clues on the following.

Prices.   A low rate of profit in a firm relative to that in the industry, for example, reveals that costs are too high or selling prices too low or that the plant is outdated or inefficiently run. In either case, it's possible that prices may be subject to some upward pressures. On the other hand, if the firm's low profit is caused by inadequate volume, it may be willing to negotiate a lower price to build sales.

Conversely, a too high rate of profit also has some important price implications. It may give the buyer leverage in negotiations. It can possibly be used to put the supplier on the defensive—in that he might be asked to justify the price he is quoting. This is often possible if your requirements are a substantial part of the firm's capacity.

Inventories. The size of vendors' inventories relative to sales can also give some pretty important clues. For example, a slow inventory turnover rate relative to that of others in the same industry might indicate that a vendor is overstocked and hence vulnerable to bargaining pressure.

The inventory figures for companies with few product lines can also give some clues as to how fast a buyer can get delivery in a pinch. If figures are broken down into raw materials, goods in process, and finished goods, the buyer may get some idea of any inventory imbalance and distress merchandise on suppliers' shelves.

How inventories are valued is also an important consideration. Whether last-in, first-out (LIFO), first-in, first-out (FIFO), or average cost methods are used can make a big difference in inventories, cost of goods sold, and profits (see page 204).

Reliability. The financial condition of a vendor affects quality, delivery, and (as noted above) prices. The solvent firm is more likely to meet its contractual obligations than the firm that is bordering on bankruptcy. A vendor that is one step ahead of the bill collector is more likely to cut corners on quality, figuring that only by squeezing nickels and dimes can he stay in business. For this type of seller, it's the creditors rather than the buyers who must get top priority.

Stockholders, of course, have an equally important interest in full financial disclosure. It's virtually impossible to apply one of the tests of business efficiency—return on capital—when companies pass over in silence such key variables as true earnings, the method of evaluating assets, performance of subsidiaries, and executive compensation. When such conditions prevail, it is also impossible to make accurate comparisons of the performance of different companies in the same industry. In short, without proper and honest disclosure, there is no way of determining which company's shares are most attractive.

Meanwhile, Uncle Sam has also become increasingly interested in setting up uniform standards—not only as a means of safeguarding the investor but also because the government is the country's largest single purchaser. With federal government outlays running over $100 billion a year (10 percent of the GNP), there's increasing pressure to know the cost-price-profit position of its suppliers. Thus in 1970 the Controller General of the Government Accounting Office (GAO) pressed for uniform standards which would give federal purchasing specialists a firmer fix on supplier costs and strengthen their hand in pricing and administering negotiated contracts. He pointed out that close to 90 percent of all government buying was then done on a negotiated basis.

One big beef at that time: The then existing accounting principles permitted contractors to classify direct and indirect costs in many different ways. Nor were uniform rules available for allocating costs between government and commercial work. At that time contractors could also choose among varying methods for charging off such costs as depreciation, inventory, lease financing, independent research, and general and administrative expenses.

To end much of this, the GAO has long aimed at establishing standards of consistency and disclosure. A standard of consistency is particularly important. This would, for example, forbid a contractor to change, during the life of a contract, from accounting practices approved at the outset, without notifying the government.

Another factor in the push toward uniform defense contracts has been the feeling on the part of some people in the government that Uncle Sam is being taken. Indeed, one outspoken Navy man, Vice Admiral Hyman C. Rickover, said in early 1970 that the government could save at least $2 billion a year by requiring defense contractors to follow uniform cost accounting standards.

He added that the lack of such uniform standards at a time when 100 companies provide two-thirds of the $40 billion worth of goods purchased by the Pentagon is "the most serious deficiency in defense procurement." Rickover further charged that the defense contractors made exorbitant profits but hid them as costs through accounting maneuvers.

Replying to critics' charges that he knew nothing about accounting, he said: "You don't have to be a hen to smell a rotten egg. Nobody in government knows what profits are being made by defense contractors. I don't know, Congress doesn't know, the Pentagon doesn't know—only the contractors."

Rickover at the time based his estimates of a $2-billion savings on his own experience as the Navy's chief builder of nuclear submarines: "In my dealings with contractors, whenever I get into negotiations and have the time to do it, I can save 5 to 10 percent. On a $40-billion procurement budget that's a savings of at least $2 billion."

Hiding the Facts

While the attitude toward full disclosure is changing (as noted above), it should also be emphasized that the completely candid report is still a rarity. There are still plenty of incentives for hiding pertinent information. The usual assumption is that facts are hidden to fool the tax collector. But this just doesn't hold water. Multimillion-dollar firms find it increasingly hard to keep secrets from the prying eyes of the Bureau of Internal Revenue. Indeed, if firms would disclose the same figures in their annual reports that they declare on their tax returns, they would be taking a giant step toward full disclosure.

A good example of this dual type of reporting will be brought out in Chapter 13, which describes how many firms have switched away from accelerated depreciation for stockholder reporting—while keeping it for tax purposes. Thus many corporations today opt for the best of two worlds—showing big profits to stockholders and smaller ones to Uncle Sam's tax collectors.

To justify a fuzzy reporting approach, some executives argue that company financial affairs must be hidden from the inquiring gaze of trade unions. But that contention has lost much of its validity these days, when labor representatives sit on the boards of many large corporations. It is probably safe to say that today, unions know more about the true state of affairs than small shareholders do.

Another excuse against full disclosure is that it would injure the competitive position of the company. The experience of many firms would seem to refute this. But the belief still persists—with management fearing that being frank and open would put them at a disadvantage against more secretive rivals.

A few companies even admit that they are worried about their own stockholders. Fuller disclosure, they argue, would expose management to intense shareholder pressure for higher dividends, making it harder and harder to set aside adequate sums for reserves, research, and future expansion. To avoid such pressures, not only do some reports remain excessively vague, but annual meetings are often held at inconvenient times in remote places to keep attendance low—thus lessening the chance that embarrassing questions will be asked.

Finally, some companies "hide" the facts simply because they don't have the information for full disclosure. There's the classic story of a chemical firm which made the decision to consolidate all its subsidiaries—but had to postpone the move for two years while it gathered together all the pertinent information on sales and earnings.

Assuming for a moment that the decision to hide pertinent facts is a deliberate one, how can this goal be accomplished? The list of ploys is long—and in many ways the one to be used depends on the particular type of industry or operation. Nevertheless, some generalized comment on the approaches can be made.

One of the most common ploys is failure to consolidate sales and earnings of subsidiaries with those of the parent company. Thus as late as 1968 some 15 percent of the profit figures in *Fortune*'s 500 list still bore the warning footnote: "Not fully consolidated" or "Parent-company figures only." In each case, without a consolidated balance sheet, the net figures were virtually meaningless.

Nonconsolidation can often hide losses by fictitious sales to subsidiaries. One spectacular case a few years ago showed that management had juggled the books to the extent that the firm, capitalized at only $20 million, had piled up debts of $320 million before it went broke.

Accumulation of hidden reserves is another popular distortion. This often enables management to keep unaccounted-for financial resources to meet later emergencies—such as making dividend payments in bad years. One way of hiding such reserves is by deliberately undervaluing assets. A firm, for example, may carry building and property on the balance sheet at a nominal $1 value. Then, too, some outfits hold substantial real estate or stock as investments, but one would never know from their balance sheets how much these assets have appreciated in value over the years.

Other motives can produce still other ways of doctoring up financial statements. For competitive reasons one firm may fail to reveal part ownership in another. Then, too, a popular way of concealing assets is to undervalue inventories. Still other firms in financial hot water may elect to put a high value on a substantial volume of unsalable goods—or even to treat nonrecurrent income from sales of assets as if it were part of normal operating profit.

The list could go on and on. But the above should be sufficient to convince the analyst that things are not always what they seem to be. Nor are these tactics always illegal or shadowy. In many cases such moves have legal sanction. Thus as pointed out above, many firms, for reporting (not tax) purposes, recently switched—with Uncle Sam's blessings—from accelerated depreciation in order to make profits look larger to stockholders.

Financial Statement Analysis

In any event, a thorough knowledge of financial statements has become a must for both analysts and people who deal with the firms in question. Indeed, even with the possibility of distortion discussed above, there is a wealth of information tucked away in operating statements—including valuable clues on financial strength, reliability, etc.

With this in mind, a sample balance sheet and income statement (sometimes referred to as a "profit and loss statement") are presented in Table 10-1. They are oversimplified—with many key items omitted—to emphasize some of the more commonly used financial yardsticks for evaluating a firm's overall business health.

Note first that the balance sheet gives the condition of the XYZ Corp. as it was on December 31, 1970. It is not a history of a year's operations, but rather a "snapshot" of the firm on that day. The accompanying income statement gives a summary of how the company operated over the entire year ending December 31, 1970. All the yardsticks given immediately below can be derived from the figures appearing in these two simple financial statements. (Each line in Table 10-1 is lettered to facilitate the exposition.)

Current ratio **(E/M).** This is a commonly used measure which indicates a firm's liquidity (ability to meet current obligations). Defined as the ratio of current assets to current liabilities, it can be obtained by comparing item E to item M in the balance sheet. That would be $21 million, compared with $10 million—or a ratio of 2.1:1.

Since a figure of 2:1 is generally considered satisfactory, there seems to be nothing to worry about on this score. A firm with a ratio of less than 1:1 has a real problem, according to some analysts. But it is also generally agreed that the more liquid the current assets, the less is the margin needed over current liabilities.

Quick ratio **[(A + B + C)/M].** This is another liquidity measure. It's defined as the ratio of the total of cash, current investments, and

TABLE 10-1 *Sample Financial Statements*

XYZ Corporation Balance Sheet
on December 31, 1970 (Millions of $)

ITEM	ASSETS		ITEM	LIABILITIES	
	Current Assets			*Current Liabilities*	
A	Cash....................	$ 6	J	Accounts payable.........	$ 5
B	Government securities......	5	K	Accrued taxes.............	3
C	Accounts receivable........	4	L	Accrued wages & int......	2
D	Inventories..............	6	M	Total current..........	$10
E	Total current..........	$21			
			N	*Long-term Bonds*...............15	
	Fixed Assets				
F	Building & eqpt..........	$79		*Stockholders' equity*	
G	Less accumulated		O	Preferred stock...........	12
	Depreciation...........	20	P	Common stock...........	28
H	Net fixed assets........	$59	Q	Surplus.................	15
I	*Total assets*.................	*$80*	R	*Total liab. & equity*..........	*$80*

XYZ Corporation Income Statement
Year Ended December 31, 1970

ITEM		(MIL. OF $)
S	Sales..	$21
T	Less cost of goods sold.......................................	$15
U	Gross profit..	$ 6
V	Less administrative & selling expenses..........................	$ 1
W	Less depreciation...	$ 1
X	Less interest charges..	$ 1
Y	Net profit (before taxes)......................................	$ 3
Z	Less income taxes..	$ 1.5
AA	Net profit (after taxes)..	$ 1.5

accounts receivable to total current liabilities. Inventories are eliminated in the asset section because they can't usually be converted into cash as quickly as accounts receivable or government securities. For the XYZ Corp. the ratio is determined by adding items A, B, and C and comparing them with item M. That's $15 million, compared with $10 million—or a ratio of 1.5:1. A 1:1 ratio is generally considered normal, so the XYZ Corp. also passes this test.

Sales/receivables ratio (S/C). A comparison of these items aims to show whether customers are paying their bills on time. In the above example, yearly sales are $21 million (item S from the income statement). Assuming terms of net 90 days for industry, you wouldn't want much more than $5¼ million (the equivalent of 90 days' sales) in receivables. If receivables were more, it might indicate slow payment, difficulties in collection, and general aging of debt. Since the XYZ Corp. has $4 million in accounts receivable outstanding, the firm is again in pretty good shape. Those interested in obtaining industry-by-industry yardsticks on days' sales outstanding might want to consult the periodic (quarterly) reports released by the National Credit Research Foundation. This association's figures show (1) wide variation among industries plus (2) a gradual trend over the past decade toward slower and slower payment.

Cash flow (**W + AA**). This represents the sum of net income (after taxes) plus depreciation, or how much money is flowing into corporate tills. The sum of the two is a far better indicator than profits alone of a company's ability to modernize and expand. Indeed, some analysts feel it is perhaps today's best indicator of a firm's financial health because it is little affected by shifts in depreciation methods. The depreciation segment of cash flow, for example, may tend to rise if a shift toward accelerated depreciation is effected. But any such rise will tend to be offset by a fall in the profit segment of cash flow since depreciation is considered a cost element. In short, cash flow is a more consistent measure than either depreciation or profit. For the mythical XYZ Corp., cash flow consists of items W and AA—or $1 million plus $1.5 million, or $2.5-million.

Margins (**U/S and Y/S**). A study of gross and net profits (before taxes) relative to sales also provides a lot of interesting information about the vendor. Profits are the lifeblood of the firm, and if they're down, it may spell trouble. And as pointed out earlier, a look at profits also gives some guidelines for price negotiations. If a vendor's profits are above the industry average, then some questioning on his listed prices might be in order.

Also of particular interest as far as prices are concerned is the gap between gross profit and net profit. This gives information on operating expenses—a factor which often weighs heavily in vendor pricing decisions. In the case of the XYZ Corp., gross profit margin (item U divided by item S, or 6/21) comes out to 28.6 percent. The net profit margin before taxes (item Y divided by item S, or 3/21) comes out to 14.3 percent. That's a gap of more than 14 percent—though, of course, whether it is too large or too small again depends upon the industry. If, for example, the gap is much larger than the industry average, it could mean that the vendor is padding his payroll or perhaps that he has a relatively inefficient administrative and sales setup.

Inventory turnover (**T/D**).   This figure (cost of goods sold divided by inventories) is designed to show how successful the firm has been in controlling total inventories. It also can give some hints on the quality of inventories. Thus, if the ratio is relatively low, it could mean that the firm has a lot of obsolete materials or unsalable finished goods on hand—and hence may be in some sort of trouble. It follows, then, that a high turnover rate is always preferable to a lower one.

Other things being equal, the more rapid the inventory turnover, (1) the greater the profit possibilities of the firm, (2) the smaller the possibilities of inventory obsolescence, and (3) the lower the current ratio upon which the firm can safely operate.

But since these figures vary sharply by industry (see Figure 10-1), no rule of thumb can be given as to what constitutes a high or low rate. Generally speaking, a firm's turnover rate must be compared with that of its industry. In the case of the XYZ Corp., the T/D ratio comes out to $15/$6, which boils down to a turnover rate of 2½ times a year.

Price/earnings ratio.   This is calculated by dividing the market value of a stock (not shown in Table 10-1) by the firm's annual net earnings (AA). In other words, it is the rate at which the securities

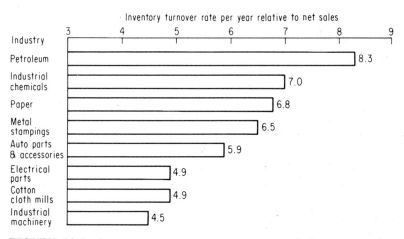

FIGURE 10-1   *Inventory turnover rates per year relative to net sales* (*Dun & Bradstreet five-year average*).

market capitalizes the earnings of the company. Thus a stock that is earning $5 per share and sells for $100 has a P/E ratio of 20:1.

This is perhaps the most widely recognized yardstick of a stock's attraction—for other things being equal, the lower the P/E ratio, the more attractive the security. Many use it to determine whether or not to buy a certain security. But like any rigid yardstick, the P/E ratio must be applied with caution. It should by no means be the only—or even the deciding—factor in a decision on whether or not to buy or sell a certain security. Clearly, any blind selling of growth stocks when the P/E ratio reaches even as high as 30:1 can be disastrous. The history of such growth companies as IBM and Xerox proves this point beyond a shadow of a doubt.

Nevertheless, some broad-gauge P/E rules are valid for decision making. Thus a company that consistently has increased its profits and net per share by 10 or 15 percent annually should generally carry a higher P/E ratio than a sluggish, slow-growing outfit. The reason, of course, is that investors in the first instance are considering the company's greater growth potential—and are willing to pay a price for this potential.

But again some caution is still advisable. Thus stock market fashions may change—and an industry that today attracts a high price because of its glamour may not do the same tomorrow. Thus in the early 1960s boating, bowling, computers, and conglomerates were all glamour issues with high P/E ratios. But by 1970 all lost favor with investors—with some P/E ratios in these areas dropping from over 40:1 and 50:1 to under 10:1—even in cases where the companies involved lived up to their growth expectations.

Inventory Evaluation

The entire problem of inventories—discussed briefly above under "Inventory Turnover"—requires further comment. Considerable emphasis is put on this area by financial analysts because:

1. Material costs almost always add up to a large—sometimes the largest—expense item on the income statement.

2. Inventories are usually a substantial part of the net assets of a company, and hence they figure importantly in any liquidity evaluation.

3. Inventories can be, and usually are, evaluated in many different

ways. For example, whether a firm uses LIFO, FIFO, average costs, etc., can make a big difference in the inventory figures and in the firm's overall profit position.

FIFO (first-in, first-out) assumes that all withdrawals from inventory to the production line have been made from the earliest purchases—at then existing prices. Thus in times of rising prices, cost of goods sold tends to be lower and profits higher than they would have been if the material leaving inventory had been costed at current higher market prices. At the same time, inventories shown on the balance sheet at the end of the period (since they represent the cost of the most recent purchases) more closely represent current market value of inventory assets.

LIFO (last-in, first-out), on the other hand, assumes all withdrawals to have been made from the most recent purchases. It follows, then, that in times of rising prices, the cost of goods sold will be somewhat bigger and profits lower than under FIFO. At the same time, inventories shown on the balance sheet at the end of the period—since they represent the lower purchase prices of earlier periods—will tend to understate the current replacement value of inventory assets.

There are pros and cons to each technique. Moreover, there are other techniques also permitted under current accounting rules. Indeed, the whole question of inventory valuation can be illustrated by a simple example. Here's how a merchandising firm (no labor or fabricating expense needed to convert from purchases to sales) might show up, using the three major types of inventory accounting.

Assume the following: The firm has no stocks at the beginning of the year, but makes a purchase of 50 units at $1 each in January and another purchase of 50 units at $1.50 each later in the year. In other words, the actual cost of the year's purchases comes to $125. Further assume that 50 units are sold during the year at $3 each—for a sales volume of $150.

Note from Table 10-2 that under FIFO, you would charge out as cost the price of the first 50 items purchased (line 3). In this case, it would be 50 units at $1, or a total cost of $50 for the 50 items sold.

Under the LIFO technique you would charge off as cost the price of the last 50 inventory items purchased. In this case, it would be $1.50 per unit, or a total cost of $75 for the 50 items sold.

Under the average cost method, divide the 100 units purchased during

TABLE 10-2   *Comparison of Inventory Methods**

	FIFO	LIFO	Avg. cost
1. Purchases during year..................... 125	125	125.00
2. Sales.................................... 150	150	150.00
3. Cost of sales............................. 50	75	62.50
4. Gross profit (2 minus 3).....................100	75	87.50
5. Inventory, end of year (1 minus 3)........... 75	50	62.50

* Under any of the above methods it's still possible for the current market value to be below the resulting inventory value. If such is the case, inventories are written down to market value to conform to the "lower of cost or market" rule. This is used to prevent overvaluation of inventories on financial statements.

the year into the total purchase cost ($125), and you come up with a per-unit cost of $1.25, or a total cost of $62.50 for the 50 units sold.

Note that LIFO shows the smallest end-of-year inventory (line 5) and the smallest profit (line 4). This is always the case in times of rising prices, as assumed in the above example. Under such conditions, LIFO generally has inventories which are valued at less than their current market value.

Many companies prefer this method during periods of rising prices because the resulting lower profits mean lower taxes. Therefore, over the past decade, when prices were generally rising, it's no surprise that many firms switched to this method of valuing inventories. But remember that the government won't allow you to switch back.

Leasing

This is another burgeoning area of corporate finance and hence financial statistics. The phenomenal growth of this technique in recent years attests to its dollars-and-cents payoff under certain conditions. But as is true of anything else in the financial arena, there are pitfalls as well as advantages—and more often than not, complex tax considerations ultimately determine whether to go the buy or lease route.

Such tax considerations are, of course, beyond the scope of this book. Nevertheless, there are some basic facts which should be considered by anyone faced with making the lease-or-buy decision. As a starter, the pluses of leasing always loom larger if your company is short on cash but long on expansion—even though in plain dollars-and-cents terms leasing will always cost you more than outright purchase.

That's not to say that if you have the money to buy you should put leasing on your list of don'ts. It's a flexible tool every firm should keep in its bag of procurement tricks—especially for emergencies.

Here are some examples of situations in which leasing equipment might be better than outright buying:

▪ A small company with little extra cash in the till is opening up a new plant. By leasing, the company can get around the large capital outlays that would be necessary to buy the equipment. It should be remembered that going to a bank and borrowing would still be cheaper. But conditions in the money market won't always permit this. In other words, bank credit isn't always available.

▪ An organization needs to expand production rapidly to meet an influx of new business—but probably the need is only temporary. Here purchasing equipment that must sit idle later on and take up storage space may be more costly than leasing.

▪ Finally, a small or newly formed company—not well equipped to perform maintenance, administrative, and service functions efficiently— may save money through leasing. One of the big advantages of leasing is that all problems involved in outright ownership (including arranging for insurance) are taken care of by the lessor.

If you do lease, many advise going directly to the manufacturer because it's cheaper. Another key point: Read the lease carefully because normally there are severe penalty clauses for cancellation.

Another key advantage claimed for leasing is that the company gets new equipment every three or four years, which boosts employee morale. Another plus claimed is that leasing provides flexibility. Most lessors, in order to enhance this advantage, offer a variety of leasing plans.

For example, one big office equipment firm in 1970 offered two leasing plans for its electric typewriters. Under its equity lease plan, the annual cost was $284.40, based on a three-year lease. However, after three years the machine could be purchased by the user for $48, or the lease could be extended for $80 yearly (including full maintenance).

Under its nonequity lease plan, the annual cost was $208.80, based on a four-year lease with no purchase option. The firm also had a straight rental fee for the machine of $270 a year. As a rough cost comparison, this same typewriter could be purchased outright for $480, plus $42 for a maintenance agreement.

Since leasing involves both a decision to invest and a decision to borrow simultaneously, sifting out the real costs may involve fairly sophisticated analysis. Thus, it's a good idea to bring your accountants in before any final decision is made. For instance, one firm looked into leasing a couple of years ago but decided it was not of interest, mainly because their accounting and tax procedures were set up to handle ownership of equipment.

A few simple questions and answers can best sum up the basic plus and minus points of equipment leasing:

- Do leases stretch limited capital?

Generally, yes. If you need additional machinery or equipment but don't have the cash for a down payment, you can usually acquire the item you need through a lease with no money down. Beware, though, of stretching your capital too far. Leasing does create debt obligations, and lease charges need to be taken into account along with other cash needs. If your major lender feels that a long-term lease strains your ability to repay, he may be less inclined to make loans for other purposes.

- Does a long-term lease increase annual cash flow?

Not always. In the year of acquisition, the lease charge will probably require less cash than a down payment. Later, though, charges will exceed financing costs. An operator with profitable alternative uses for his money may increase his cash flow over time, but this is not assured.

- Will a lease lower total costs?

No. Leasing machinery or equipment is more expensive than buying it outright. Lease charges cover not only equipment costs and interest charges but also the lessor's operating costs and his profit.

- Will a lease reduce taxes?

Yes, but unless you are in a rather high income bracket, any tax savings will be small. And the contract must meet IRS leasing specifications, or only part of the annual charge will be allowed as taxable expense. If your lease allows a full write-off of leasing charges, your taxes could be reduced or deferred to later years. Leases also permit recovery of equipment cost in fewer years than depreciation. Thus, your annual charges may exceed normal depreciation and interest charges.

- Do leases always reduce the risk of obsolescence?

No. Machinery can become obsolete regardless of whether you own or lease it. And a lease still binds you to an annual outlay for the period of the contract. However, writing off the cost over a shorter period of time may give you more flexibility.

Leasing works out best for the firm that has a high rate of return on its money and can free additional cash through leasing. It also pays for the outfit that uses machinery or equipment heavily. Then leasing allows the firm to relate tax-deductible expense more closely to the useful life of an item.

If leasing looks like a route you should follow, remember: It's best to do some detailed figuring with your lender and tax consultants before signing any agreement. And be sure to check both the short-run and long-run effects of the lease. The profit picture over a period of years may be quite different from that for just one year.

Before leaving the subject, it may also be well to distinguish between renting and leasing. The two terms are often confused. The key distinction is the time element. Rentals generally run for a short period of time—say, a day, a week, or a month. Moreover, the customer can return the product whenever he wants to and pays only for the time he has used it. Leasing, on the other hand, usually involves a time period of a year or more—and involves a contract for an agreed-upon length of time.

One last comment on the growing popularity of both renting and leasing in the consumer field. More and more families are finding it worthwhile to forgo outright purchase. As a result, they now rent such diverse products as mink coats, wheelchairs, power saws, and even a complete table service for a special dinner party.

Aside from the advantages for industrial customers noted above, the increasing popularity of renting and leasing in the consumer area can be attributed to several special factors. One of the most important of these is the high interest rates prevailing today, which discourage people from borrowing to make major purchases.

Another factor is the still-accelerating do-it-yourself trend—particularly among the growing army of suburbanites. This has created a brisk demand for rented tools. Why purchase an electric saw or fancy drill when you can rent one at a much lower cost for the occasional job that may crop up in the home?

The increasing amount of leisure time is still another key force. More time for recreation and entertainment has meant more demand for rented tents, boats, skis, etc. Then finally there's an increasingly mobile population to reckon with. Why accumulate roomloads of equipment that has to be packed, shipped, and then unpacked every time a family pulls up stakes?

Limitations of Financial Statistics

In some instances it is not dishonesty or legal sanction, but simply the rules of the game, that make financial analysis so difficult and tricky. A perfect example is the way different forms of raising capital are treated from a tax standpoint. Thus, depending upon whether money is raised in bond or equity markets, the net profit may be either smaller or larger.

A simple example will suffice to illustrate this point. Assume two identical corporations—one financed by $100,000 in 6 percent bonds, and the other by $100,000 in stock. Further assume that each firm earns $10,000 after all costs (except bond interest) are covered. It then follows that the corporation taking the bond route will show a book profit of only $4,000 because bondholders' interest eats up $6,000 of the $10,000 operative profit. On the other hand, the corporation going the equity route will show the full $10,000 profit on its books.

The question then arises as to whether the stock-financed firm is a better-run corporation than the bond-financed one because it shows a higher "book" profit. The answer, of course, is "no." The only difference between these two companies stems from different accounting rules. The law permits firms to deduct bond interest as a cost in calculating profits for income tax liabilities—but makes no such provision for the implicit return on stockholders' investment.

Actually, the accounting profit routes enumerated above are just two of many—neither of which can alone provide the definitive answer to every business problem. As a matter of fact, there are literally dozens of different profit yardsticks that one might want to use.

The entire stable of such earnings measures would include accounting profit, economic profit, real profit, time-adjusted profit, contribution margin, controllable margin, operating profit, long-run profit, short-run profit, average profit, and marginal profit. Moreover, "decision theory" has introduced such new concepts as maximum profits and profits adjusted for the utility of money to the decision maker. Then, too, there are combinations of the above such as long-run real-time adjusted-marginal profits. And finally there are ratios—profits expressed as a percent of (1) net or gross assets, (2) net worth, (3) number of shares outstanding, (4) value of shares outstanding, and (5) sales.

The basic point is this: The profitability of a proposed decision can vary substantially—depending upon the measure of profitability used. Put still another way, what could be profitable using one concept could well turn out to be unprofitable under another concept.

Thus many companies find the use of any one concept too narrow and confining to meet all their needs. In these cases, they generally keep several sets of books or worksheets. One is generally earmarked for tax purpose, another for stockholders, and one or more for other decision-making purposes.

But even with all these different sets of books, it is often difficult to pinpoint true costs or profits. Interest charges are a case in point. Any accountant, for example, will tell you that inventory financed internally does not carry any interest charges. Yet it is clear that if carrying these same inventories required borrowing, then the subsequent interest charges would be considered a cost of doing business.

Two reasons are often advanced for this inconsistent approach: (1) Tax laws are so written that only "outside" interest is deductible, and (2) measuring "inside" interest costs is extremely difficult. Thus in the latter case, one can use the normal rate of return enjoyed by the firm, the outside rate for similar loans, or the interest that could be earned if the money were kept in a bank. The lack of any unique interest figure is also probably the reason why tax law has never recognized this true cost as an accounting cost.

Yet any businessman worth his salt obviously has to consider the cost of capital when determining optimum inventory levels—whether that capital is externally or internally generated. Note, too, that these same interest cost considerations are equally important when considering whether to buy or not to buy a new piece of equipment. They should also figure prominently in the lease-or-buy decision discussed in the previous section.

The basic problem of capitalization versus expensing also involves different concepts of costs and profits. Talks with accountants, for example, suggest that treating items such as advertising as period costs (instead of as a capital asset) stems from the fact that it is not easy to determine how much of the advertising outlay can be allocated over any given period. Who, for example, is to say how much sales changed in one period because of an advertising outlay in a previous period? Thus accountants generally prefer to expense the "hard" figure used in the invoice. It is a lot easier than guessing at the right figure. But this doesn't make it correct or useful for management decision-making purposes.

In another sphere—even where costing is allocated correctly—strict adherence to accounting methodology can be misleading. That's because accounting records indicate what has already happened—not what should have happened. An example again can best illustrate this point.

This time consider a petroleum refiner who has the option of producing fuel oil or gasoline from his basic raw material. Assume that he opts for producing fuel oil at a profit of, say, $15 per barrel. If he can sell 1,000 barrels, his profit would be $15,000. On the other hand, he could have converted all his raw petroleum into gasoline, which would have yielded 1,000 barrels with a $20-per-barrel profit. Thus his gasoline profit would have been $20,000. In short, his decision to produce fuel oil was a bad one—resulting in a $5,000 drop in his profit potential.

But a look at the profit and loss statement can in no way give any inkling of this poor decision. All the statement would show is that the company made a fair-sized $15,000 profit. It doesn't tell what might have happened—and hence is a poor yardstick of managerial effectiveness and savvy.

There is one final example concerning the inadequacy and limitations of financial statistics, this one pointing up the fact that an unsophisticated look at the accounting figures can well lead to the wrong decision. The illustration this time around involves a two-product company where one of the products shows a profit, and the other one a loss. The big question facing management is, Should it drop the unprofitable line?

Here are the details (they're also summarized in Table 10-3). Product X has been losing $3,000 a year because its costs ($16,000 variable and $7,000 fixed) have exceeded the $20,000 sales total. Product Y, on the other hand, has sales of $100,000 and costs of $96,000 ($56,000 variable and $40,000 fixed)—making for a profit of $4,000. For the company as a whole, then, only the profitability of product Y has seemingly enabled it to come up with a small consolidated $1,000 earnings gain.

Next assume that an unsophisticated management decides to eliminate product X in order to beef up profits. Here's what actually happens: The $7,000 in fixed expenses previously allocated to product

TABLE 10-3   *The Dangers of Product Elimination*

A. *Before elimination of product X*

PRODUCT X			PRODUCT Y		
Costs:			Costs:		
Fixed.........	$ 7,000		Fixed...........	$40,000	
Variable.......	16,000		Variable........	56,000	
		$23,000			$ 96,000
Sales...................		$20,000	Sales.....................		$100,000
Loss....................		$ 3,000	Profit....................		$ 4,000

Consolidated profit = $1,000

B. *After elimination of product X*

PRODUCT Y		
Costs:		
Fixed........	$47,000	
Variable......	56,000	
		$103,000
Sales..................		$100,000
Loss...................		$ 3,000

X must now be allocated to product Y. As a result, total expenses for product Y rise to $103,000—topping sales by $3,000. In short, the decision to scrap product X has actually lowered consolidated profits to the point where the firm is now suffering a $3,000 loss, rather than raising them from the previous low $1,000 level (see Table 10-3).

Management's folly in this case was, of course, its failure to take into account the reallocation of fixed costs. The mistake could easily have been avoided by computing the so-called contribution margin—the difference between sales revenue and variable costs. In the above example, the contribution margin of product X was $4,000 ($20,000 in sales less $16,000 in variable costs). That's a positive figure and hence its survival should have been justified. As one analyst puts it, "Decisions of this sort must consider only variable costs—fixed costs in this case are essentially irrelevant."

All the above is in no way meant to demean the value of accounting records. All that is implied here is that accounting records aren't the beginning and the end of everything. They're valuable, yes, but they also have their limitations. And to know what these limitations are can permit more meaningful use of the available statistical intelligence.

Other Financial Caveats

Aside from the limitations enumerated above, there is always the possibility of reading into financial statistics meanings which are simply not warranted. Thus a few final warnings on overall interpretation may be in order. Specifically, always keep out a weather eye for the following.

Reliability.   One major question all analysts ask is, Are the reports audited by an accredited accounting firm? Generally speaking, an audited report puts the reputation of the accountant behind the statement. It's not likely that these auditors would jeopardize their whole practice to cover up for just one firm. It's a pretty good rule to accept an audited report rather than one that has not been audited.

Comparability.   Don't automatically put one firm's inventory or profit figures up against another's, for individual companies often have different methods of valuing inventories, taking depreciation, etc. And differences here lead to differences in profits and profit margins. The thing to remember is to check the accounting methods which underlie each figure in the balance sheet and income statement and to read carefully the footnotes to statements.

Industry differences.   Don't compare profit margins blindly. Each industry has its own norm, and it's usually much more important to check against industry averages than any national average.

Short-run problems.   Often a strike or other unusual event can distort operating results because what happened in that year is not typical. A recent steel situation provides a good example of this. If you were to look at the second quarter of 1968 alone, a steel company's earnings would look much rosier than would be warranted because of hedge buying. And if you were to look at the third quarter alone, the picture would be too bearish because sales and profit declines reflect the ensuing inventory cutbacks. A look at the entire year in this case would give the better picture.

Most analysts, however, prefer to look over a firm's history for several years. For in addition to washing out one-shot affairs, this helps give a clearer picture of the trend over time—and therefore of the basic health of a particular firm.

ELEVEN

Hidden Influences

There's more to a statistic than meets the eye. An obvious statement? Yes. But sometimes the hidden aspects of the number or numbers in question aren't nearly so obvious.

More to the point: The figures presented can be accurate, reliable, and free of distortion—and in general can represent what they purport to represent. Nevertheless, they can be misleading if used as a springboard for making important policy decisions.

That's because there is the very real possibility that hidden factors may be influencing the magnitude of the number under study. Thus, dollar sales may be rising smartly—but much of the advance may be due to price rather than to any increase in the number of units sold. Hence, if these dollar sales gains alone were used as a basis for capacity expansion, they could conceivably lead to overestimation of future needs and possible future glut.

In short, the available statistics must be carefully monitored to see whether they are directly applicable to the problem or situation under study. In many cases when they are not, adjustments can be made to make them applicable. Thus in the case of dollar sales figures cited above, the price influence can be removed—thereby making the sales data suitable for capital expansion analysis.

There's an added dividend to this approach to sales statistics. In addition to zeroing in on physical volume, it forces the analyst to take another good, hard look at the independent role played by price—a variable sometimes lost in the shuffle. Since the latter factor must be isolated to shift from value to volume, it presents a tailor-made opportunity to (1) explore statistically how price in itself can influence sales and (2) compare your own company's price performance with that of the industry or even the overall economy. Such price data may also be used as a measure of your own cost performance—the extent of rise or fall in prices, indicating how well your firm has been able to hold the cost line.

All the above emphasis on physical volume and price by no means implies that the dollar figures should be ignored in statistical analysis. Far from it. There are occasions, for example, when it may well be the dollar rather than the volume data that are called for. Again, as in all statistical work, the problem at hand determines the choice of variables.

Harking back to the use of dollar sales—such figures, while clearly unsuitable for capital planning, can become an integral part of any overall master plan involving future profits and cash flows. It is also this dollar figure which must be used when comparing a company's sales performance with that of individual competitors or the overall industry.

So far, only prices have been singled out as a hidden influence. To be sure, they are one of the more common factors, but hardly the only one. It is the intention here to put the spotlight on several hard-to-see statistical influences that normally crop up in the type of data analyzed by the typical industrial firm—all in the hope that this will lead to a more critical evaluation and analysis. With this in mind, the following areas will be discussed below:

1. Prices—isolating the quantity and price components of a dollar total and where to use each

2. Quality—distinguishing between what part of a price change is true price and what part is due to a shift in quality

3. Seasonal factors—pinpointing how much of a data change is due to an underlying trend and how much to seasonal swings

4. Noneconomic factors—separating out one-shot influences such as strikes, which can often distort the underlying trend

5. Productivity—determining how much of a change in labor requirements is due to basic demand shifts and how much to shifts in productivity

In addition, recognition will be given to the fact that many hidden influences are often combined in a set of raw data. Thus it is possible to start with a value series, remove true price and quality in two separate steps, and end up with a real, or deflated, series.

Similarly, seasonal, price, and sometimes one-shot factors must often be removed from a given series. Thus sales in a given month can be simultaneously influenced by price shifts, a seasonal shift, and a major strike in a customer industry. To ignore one or all of these is to open the way to possible distortion and mistakes.

Value, Volume, and Price

As pointed out above, many times the basic data are presented in value terms, that is, as a combination of the number of units multiplied by the unit prices. If, however, the situation calls for an analysis or projection in physical terms, then the value data must be adjusted to eliminate the unwanted price factor. Traditionally, this is accomplished by the process of deflation.

The term "deflation" is particularly appropriate, for essentially what the analyst does is to take the "air" out of the figures—i.e., he transforms the data into what the statisticians call "real" terms.

The technique, while now widely used in the processing of business data, received its start in the general economic sphere. In periods of price inflation, for example, many of the apparent gains in GNP and income are basically illusory. Thus if both income and price go up 5 percent in a given year, the net gain to the consumer or businessman is nil. The 5 percent extra income buys no more real production or services.

The recent period of inflation which the United States went through illustrates this effect most clearly. For many Americans the apparent sharp economic gain of 1969 turned out to be an illusion created by inflation. While the GNP grew by a reported 7.5 percent in current dollar terms, a 4.7 percent price advance held the real or physical gain to only 2.8 percent.

Over the same period of time the average weekly earnings of a manufacturing industry worker increased by $7 or some 5.7 percent. But because consumer prices rose 5.4 percent, his purchasing power remained virtually unchanged. And for those with fixed incomes the effects were even more crushing. Pensioners, for example, suffered a loss in real income equal to the entire rise in consumer prices—5.4 percent.

The effects on corporate profits were no better. Although earnings were up close to 15 percent, if invested in new equipment they would buy little more than at the beginning of 1967. If invested in new factories, where inflation hit hardest, they probably would have bought considerably less than that.

In short, when analyzing general economic data, there is an urgent need to deflate figures to arrive at real gains. The same can be said about individual company data. Thus for the firm as well as for the economy, capital expansion must be analyzed in terms of real or physical requirements. Similarly, purchases of materials and parts depend on the physical volume of production, not on their price-inflated dollar value.

Without the process of deflation all this would be impossible. Happily, the statistical techniques for accomplishing this are basically simple. It can usually be done using a calculator or computer relatively quickly and easily—provided, of course, one has a reasonably good idea of how fast or slowly prices are rising.

On the latter score, there is plenty of price information available—both from company sources and from Uncle Sam. The government, for example, has a wholesale price index made up of over two thousand industrial prices—presented separately and in various combinations. In other words, chances are there's a pretty good price index available to the researcher if he wants to filter the price effect out of a particular series being analyzed.

The actual mathematics of deflation is quite simple. All it involves is taking the value measure and dividing it by price. The result is the quantitative, or real, measure.

This can best be illustrated by first considering the definition of value: Value (V) equals price per unit (P) multiplied by the number or quantity of units (Q). Specifically, $V = P \times Q$.

If you then divide value $(P \times Q)$ by price (P), the P's cancel out, leaving you with an estimate of quantity (Q).

In the real business world often only the value of a series and an approximation of prices (a price index) are available. (The reader is referred to Chapter 6 for a more complete discussion on the construction and use of index numbers.) In this typical case, dollar magnitudes (estimates of $P \times Q$) are divided by price indices (estimates of P). The result is a series in real or physical terms with the price effect factored out.

An example can best illustrate when and how the approach might be used. Assume that an analyst in the automobile industry wants to correlate company sales with consumer income. If unit sales are being used, then consumer income must also be on a volume basis. In this case, it is likely that he would divide the consumer-income dollar series by a measure of price changes in consumer income—more specifically, the consumer price index—thereby getting consumer income with the price effect factored out. This is usually referred to as income in *real* terms or income in *constant dollars*.

Some firms, in order to deflate their own product mix, actually construct their own tailor-made price indices. They then, as described above, divide their dollar magnitudes by their own price index, again coming up with an estimate in terms of constant dollars.

Where prices tend to change violently, the deflation of the data is almost a necessity if a satisfactory forecasting relationship is to be found. This can best be illustrated by a simple example. Assume that widget sales have been 100, 200, 300, and 400 units over the past four years. Further assume that the widgets were originally priced at $1 per unit but that because of a severe supply shortage, they went up to $2 in

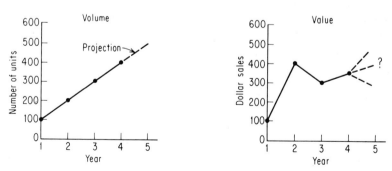

FIGURE 11-1 *Volume versus value relationships.*

the second year. They fell back to their original level during the third year as new suppliers came into the market, and they fell an additional 20 percent in the fourth year.

The problem is to extrapolate sales on the basis of past trend relationships. On a purely volume basis it is easy to see that sales during the ensuing year would be 500 units (see Figure 11-1). If data, however, are in value form, the sales figures over the original four years come out as follows:

Year	Sales	
1	$100	(100 units × $1 per unit)
2	400	(200 units × $2 per unit)
3	300	(300 units × $1 per unit)
4	320	(400 units × 80¢ per unit)

With the dollar figures over the years going both up and down (see the above table), the trend is far from clear. Hence, using these value data as the basis for extrapolation is quite risky and far less reliable from a statistical point of view. In the above case it is, of course, clear that data with the price effect removed should be used.

But there are other reasons, as well, for using deflated data in forecasting. One company forecaster expresses his preference for working in terms of physical units as follows: "We would rather project volume and let top management make any estimates as to any overall change in price over the forecast period since they have a better feel for this sort of thing than we do."

There is, of course, no problem in the volume approach when the forecast is designed specifically for production control (to schedule future output) or for long-range planning (to gauge how much capacity will be needed, say, five years from now). Then volume projections are precisely what is wanted and needed.

On the other hand, if the controller wants a forecast to gauge cash inflow or credit needs, it is the dollar value of sales he is interested in. In this case, the researcher has several alternatives: (1) do the forecast, if feasible, in value terms; (2) do it in volume terms and then translate it into dollar terms by using prices supplied by his own or some other department; or (3) transmit the forecast to the department that requested it, but ask the recipient to make his own assumptions on where prices are heading.

One thing is clear, however: Mixing value and volume in the same forecast equation can lead to serious errors. In statistics such mixing of apples and oranges just isn't feasible if the analyst expects to come up with anything meaningful and reliable.

It is not a good idea, for example, to relate the dollar sales of your products to the movements in industrial production because sales is a dollar concept and industrial output is a volume concept. The proper procedure in this case is either (1) to deflate the sales series as outlined above or (2) to inflate production by multiplying by some appropriate industrial price index to obtain a value magnitude. In other words, be consistent. Use value or volume throughout. Some researchers like to try both types of projections, one based on volume relationships and the other on value relationships. Strange as it may seem, the relationships can sometimes be quite different, thereby resulting in somewhat different projections.

Quality and Price

Closely allied to price is the problem of quality. There is general agreement among statisticians that price indices should measure changes in the price for items of constant quality. Thus if quality changes, the price index should be adjusted accordingly.

More specifically, if quality improves and the asking price is unchanged, this would be considered a price-index decline because the buyer is getting more for his money. On the other hand, deteriorating quality at an unchanged asking price generally requires an upward adjustment in a price index—this time because the buyer is getting less value for his given dollar outlay.

Take the example of an earthmoving machine introduced in year 2 that can remove twice the cubic feet of earth as the original model introduced in year 1. Clearly the improved machine—since it can do twice as much work—is worth twice as much to the buyer. If the asking price remains unchanged, he is getting twice as much value for his money—which is another way of saying that he is really paying only half as much for the machine. Ergo, a 50 percent drop in the price index for earthmoving equipment would seem in order.

Next assume that the manufacturer—knowing he is giving the buyer a machine that is twice as effective—decides to double the price per unit. The buyer still has no grounds for complaint for he is getting the same volume of earth removed for his purchase dollar. In short, he is getting unchanged value or utility for his dollar outlay.

The task of the price statistician is essentially to weigh the increase in price with the increase in quality. If both are the same—as in the above simplified example—then the analyst would conclude that there has been no change in pure or real or true prices and ergo no change in the price index.

This kind of problem arises yearly in the auto field. New safety features and better performance are generally filtered out of any increase in suggested list prices to arrive at a pure price increase—that is, the price increase in a car of constant quality. One year when the new models were introduced, for example, list tags showed a 2 percent increase. But when the safety and other quality-improvement factors were filtered out, the pure price increase came to less than 1 percent.

Sometimes both quality improvement and quality deterioration must be considered. When the 1969 models were introduced, government statisticians figured that the added safety equipment, improved engines, and better ventilating systems were worth about $24. But they also noted that warranties were being halved. Regarding this latter move as a quality reduction, they figured that the car buyer was losing an almost as large $23 in reduced warranty coverage. Thus the net quality improvement was put at only $1. Put another way, virtually all the average list-price boost that year (about $70) was deemed to be pure price increase.

Strange as it may seem, there is often considerable product deterioration as well as improvement. A spokesman for Consumers Union, for example, recently noted that there are several reasons for such quality deterioration. First, he cited a recent tendency toward reduced quality control at many factories. He blamed this in part on the annual model changes for appliances and other big-ticket consumer items. His conclusion: The pressure to get the new model out in time often makes it impossible for the producer to do very much in the way of quality control.

Tough price competition was cited by this expert as still another reason for poor quality. Producers, he said, are often under severe pressure to cut back on quality to reduce price. He cited black-and-white television sets, which recently came down in price but in which the absence of horizontal control knobs and knobs to adjust brightness levels was conspicuous.

Deterioration, of course, does not always occur. Refrigerators today are much better than they were a decade ago. And such items as wringer washing machines have become much safer.

But whether it be improvement or deterioration, few statisticians will generally argue with the concept of filtering out quality change. Problems arise, however, in determining how quality should be defined and what constitutes constant quality. The problem can best be illustrated again in terms of automobiles. As the Bureau of Labor Statistics recently put it:

> The characteristics of an automobile make it possible to describe its attributes in many ways. The purpose of an automobile is basically to provide a means of transportation, but it has come to mean many more things to owners or prospective owners, depending upon needs, desires, income, environment, and training. For example, consumers value an automobile in terms of reliability, durability, convenience, safety, economy, speed, carrying capacity, comfort, appearance and prestige, to name a few. The emphasis placed on these "quality" indicators varies, however, depending upon individual needs and preferences.

Because of this varying emphasis, most statisticians have agreed to follow the physical approach in defining quality changes. Specifically, changes in physical characteristics that affect safety, performance, durability, and/or comfort and convenience are classed as quality changes. Adjustments are made for these changes so that the index will be based on prices for the same or equivalent quality.

Adjustments are also made for any accessories and equipment or differences in design which were offered as options in one year and made standard the next year, or vice versa. On the other hand, no adjustments are made for style changes when they serve no function other than to make the car appear new and/or different.

Finally, there's the problem of putting a dollar figure on these quality changes. The government, in working with autos, uses three types of approaches based on three different costs or prices.

1. Producer costs.   Auto-maker costs for labor and materials adjusted to selling-price levels are used to estimate values for structural and engineering changes whenever available. These values are preferred since they represent that portion of the total price of the fully assembled vehicle accounted for by the change in specifications.

2. Replacement-parts prices.   Prices for replacement parts, deflated for additional costs such as storing, wrapping, shipping, and extra margin usually applied to replacement parts, are used to estimate the value of quality change only when producer costs are not available.

3. *Option prices.* Market prices for factory-installed options are used when available to adjust for changes in accessories and equipment included in the price. Thus, any difference in price which occurs when equipment or accessories are made standard or optional is reflected in the index as a change in price for a car with a specified amount of equipment.

Moreover, as pointed out above, warranties (free replacement or repair of parts) are considered part of the quality a buyer is receiving in connection with his purchase. Thus adjustments are made to take account of any changes in this area. The estimated value for the change is based on costs to the producer for replacement parts and repairs.

To sum up, most price technicians attempt to filter out quality changes to provide the user with a measure of pure or real price change. Nevertheless, many people contend that changes often slip by unnoticed. These critics say that these are mostly quality improvements. Thus they estimate that most popular price indices—including the government's wholesale and cost-of-living indices—have an upward bias amounting to as much as 1 percent per year.

Taking another tack, there is a small but vociferous group of people who feel that quality should not be factored out of price indices. Their argument (again using autos) runs something along these lines: The average man knows that he pays more for a car today than he did 10 years ago. It costs more, and yet, according to the consumer price index report of a few years ago, the price of a new car in 1967 was less than that in the period 1957 to 1959. Following up on this line of thought, these critics add that if price indices adjusted for quality fall while actual seller list prices rise, many consumers may begin to have serious doubts about the accuracy of the price index—for it's hard for the average layman to reconcile higher out-of-pocket costs with an index showing a decline.

The answer to this argument is clear-cut: The fact that the population or a good part thereof believes in some myth is no reason to perpetuate it. If an effort is made on the part of the analyst or statistician, consumers can develop the conception of a changed car—namely, that there are differences in what they are getting for their money from one period of time to another. You see this even now in the case of medical care and hospital services. People know they're paying more—but they're also aware that the quality of the service they're getting is vastly improved over what it was 10 years ago.

It is often useful to integrate the price-quality breakout with both physical volume and dollar totals. An example involving machine tools (Table 11-1 and Figure 11-2) points up the relationship of each of

TABLE 11-1 *Metal-cutting Machine Tool Performance—1960–1968*

Year	Shipments Dollars	Units	Avg. cost per machine tool	Metalworking machinery price (1960 = 100)
1960	541,500,000	36,300	14,912	100.0
1961	541,250,000	30,600	17,700	100.9
1962	612,850,000	35,800	17,100	102.7
1963	638,450,000	35,900	17,780	103.0
1964	844,650,000	39,500	21,371	104.9
1965	1,022,550,000	44,700	22,867	107.9
1966	1,221,750,000	54,100	22,576	112.8
1967	1,353,200,000	55,500	24,361	117.6
1968	1,358,300,000	47,156	28,804	122.1

SOURCES: National Machine Tool Builders Association and the Bureau of Labor Statistics.

these parameters to all the others. The basic data for this analysis are contained in Table 11-1. Thus:

▪ Column 1 indicates the rather impressive (nearly threefold) advance in dollar sales over the period 1960 to 1968.

▪ Column 2 reveals that the increase in the physical number of units sold was substantially smaller over the period—in the order of only 30 percent.

▪ Column 3 (column 1 divided by column 2) indicates the near doubling of the average cost per machine tool over the period. This increase reflects both higher true prices and improved quality.

▪ Column 4 shows, however, that the true price increase (as measured by a government index) was on the order of only 22 percent—suggesting that most of the increase in unit costs was due to a major jump in quality.

All this is neatly summarized in terms of index numbers in Figure 11-2. Starting from the volume, both the true price increase and the quality improvement are added on to obtain the final result: the trend in dollar sales. This kind of graphic presentation points up that (1)

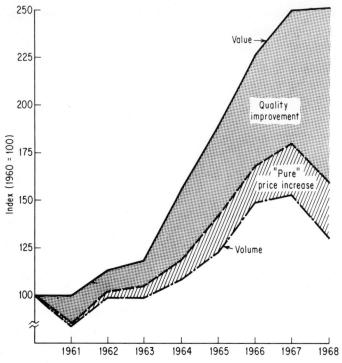

FIGURE 11-2 *Differentiating between quality and price.*

the true price rise and quality improvement together accounted for the lion's share of the dollar gain and (2) the quality-improvement factor played about four times as important a role as the true price rise.

Seasonal Distortions

Just as inflation may distort seemingly accurate raw data, so may seasonal swings. The seasonal effect on statistical data was touched upon earlier in conjunction with time-series analysis (Chapter 7). The discussion at that point centered on the several different types of forces that act upon a business or economic variable over time.

Now, however, the aim is somewhat different. As in the case of the previous discussion on prices, the current objective is to spotlight some of the practical approaches used to uncover statistical distortions.

As a point of departure, consider a comparison of raw data in a given month with similar data of the previous month. This kind of simplistic approach can work only in the absence of seasonal swings. On the other hand, whenever a seasonal effect is present (it's almost always present in sales, production, inventory, and new order data), the month-to-month comparison approach can lead to serious miscalculation and error.

A look at industrial output figures, for example, spotlights the danger in taking reported data at face value. Actual physical output (noted as unadjusted in Figure 11-3) fell sharply in July 1968.

But this in no way implied a peaking out of production. For the decline was merely the reflection of the usual summer slowdown. In fact, if you remove this seasonal effect (it's called *seasonally adjusting* the data), you see that the production trend is actually still up.

The moral is clear: When making month-to-month comparisons, data must take into account variation caused by changes in seasonal patterns if proper conclusions on trend are to be reached.

In actual practice the difficulty of getting a seasonally adjusted series differs from problem to problem. In some instances extensive calculations

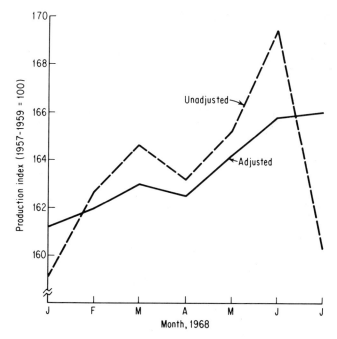

FIGURE 11-3 *The seasonal effect (Federal Reserve Board).*

of seasonals are necessary (see Chapter 7). But in other instances where the data under consideration cover an entire industry (sales, inventories, production, etc.), the work has already been done for you by Uncle Sam's statisticians. For there are thousands of useful business series available for the taking—released each month—with the seasonal factor already removed.

Even when company data are being analyzed, the government can be of help, for the seasonal adjustment factors used for its industry-wide calculations are more than likely applicable to your own firm. The only problem, then, is to determine the industry seasonal factor. This can be calculated easily by dividing the unadjusted industry figure by the adjusted industry figure.

EXAMPLE: Suppose that you make electrical machinery and note that your actual recorded output in January and July came to $970,000 and $950,000, respectively. The question is, Is business going up or down?

The first step is to determine the monthly seasonal adjustment factors from government industry-wide statistics. This means taking the government's unadjusted electrical machinery output and dividing by the government's adjusted machinery output for the months in question. Using 1968 data:

	(1) *1968* *Unadjusted* *output*	(2) *1968* *Adjusted* *output*	(3) *Seasonal* *index &* *(1)/(2)*
January	185.3	186.9	99
July	176.9	186.5	95

To seasonally adjust your own production figures, divide actual company data by the seasonal index factors calculated above. Thus your January figure would be approximately $980,000 ($970,000/99), and your July figure would be $1,000,000 ($950,000/95).

In other words, once the seasonal influence is removed, the figures show that your firm's underlying output trend is actually rising over the period in question (if you had looked only at the unadjusted data, you would have reached the opposite conclusion).

If it isn't feasible to use one of the government-supplied indices, you may want to construct your own seasonal index. This is a bit more involved, but it can be done relatively simply and cheaply with the use of a computer.

It should also be noted that seasonality applies to prices. Thus heating oil is always more expensive in winter (when demand is up) than in summer (when producers offer buyers price incentives to keep demand from drying up entirely).

Again the government has come to the rescue. The Bureau of Labor Statistics (BLS) has developed seasonal index factors for literally hundreds of different prices—thereby giving analysts a tool for removing the seasonal price factor in any month of the year. The formula to apply is

$$\frac{\text{Market price}}{\text{Seasonal factor}} = \text{adjusted price}$$

Use of adjusted rather than market prices will isolate price movements stemming from normal seasonal demand, production cycles, weather, etc., and will spotlight changes due to new market conditions and sharp shifts in demand or supply. The hypothetical example below shows how this method works on steel scrap prices.

The price of steel scrap is $26 per ton in December and $29 per ton in January. But we also know that normal seasonal influences pull up scrap prices in January. Specifically, the seasonal indices for December and January are 97.7 and 102.1, respectively.

QUESTION: How much of January's increase is due to the basic underlying trend, and how much is due to normal seasonal influence?

SOLUTION: To sort out the influences, divide the reported market price by the corresponding monthly seasonal index factors:

adjusted price per ton

$$\text{December} = \frac{\$26}{97.7} = \qquad \$26.60$$

$$\text{January} = \frac{\$29}{102.1} = \qquad \$28.40$$

The adjusted rise is $1.80 per ton (or 6.8 percent) instead of $3 (or 11.2 percent). Thus, nearly half of the jump was seasonal, with the remaining portion due to new demand factors in the steel industry. Put another way, price gains were nearly twice as strong as might be expected on the basis of normal seasonal demand.

Business analysts can use these seasonal price factors as aids in both setting pricing strategy and assessing the long-run cost of one material against that of another. By taking advantage of seasonal trends, you are playing the odds and thus over the long pull are bound to come up with cost savings.

But BLS cautions buyers to take a close look at basic changes in the market before blindly applying last year's seasonal corrections. Automobile prices, for example, fluctuate around new-model introduction time. So any change in the model introduction date will have a significant effect on the seasonal variation pattern that year.

If you keep this caveat in mind, knowledge of seasonal price can play a major role in your formulation of overall purchasing strategy. Specifically, seasonal price factors can be used for the following.

1. *Getting clues on when and when not to buy.*   These are spelled out by seasonal factors. For example, postponing or advancing a purchase for as little as one month can pay off in savings on raw materials having price swings of several percentage points over 30- to 60-day periods.

2. *Making better contract deals.*   Instead of getting vendor quotes on an average price over the period of the contract, try to negotiate for a lower price based on seasonal swings. Or, when determining the base price, check to see that it is not being pegged on quotes for periods of the year when prices are normally high.

3. *Advance planning.*   Plot price data on a seasonally adjusted basis to see whether the basic trend in a particular price is up or down. When seasonal influences are included, they distort and tend to hide the long-term price movements necessary for advance planning. For example, if two interchangeable raw materials are plotted on a seasonally adjusted basis, greater long-run upward price pressure on one may show up.

The Joint Seasonal-Price Effect

Many business and economic series are now presented in four different ways: (1) as originally reported, or raw; (2) seasonally adjusted; (3) deflated, or in constant-dollar terms; and (4) seasonally adjusted in constant dollars.

This last type of combination has become increasingly popular in recent years since it removes much of the irrelevant material that interferes with long-range planning. Widely watched United States growth statistics, for example, make use of this sort of approach. Thus, when quarterly data are released showing the economy growing at a given rate, the figure generally used is overall GNP adjusted for both seasonal variation and price changes.

More and more corporations are following the same line in analyzing their underlying business patterns and trends. For only when the price and seasonal factors are filtered out of such series as sales and orders is it possible to assess the basic growth forces behind day-to-day operations.

Many times seasonally adjusted deflated series are plotted against simple seasonally adjusted figures to point out graphically the role played by price, or the seasonally adjusted deflated series may be plotted against a simple deflated series to emphasize the seasonal influence. In short, these various combinations help the analyst separate out the roles played by the various factors that go into the building up of any reported figure.

The techniques for arriving at seasonally adjusted deflated data are essentially the same as outlined above. The only difference is that instead of dividing once (by seasonal indices to arrive at a seasonally adjusted figure—or by price indices to arrive at a deflated series), the analyst must perform two dividing operations.

Finally, a few words on annual rate may be in order—for seasonally adjusted deflated figures are often presented in terms of an annual rate. This involves nothing more than taking the reported weekly, monthly, or quarterly data and multiplying by 52, 12, and 4, respectively. This beefing up of the data facilitates up-to-date comparisons with totals of past years. It is for this reason that all GNP data are presented in terms of annual rates.

The GNP for a given quarter might show an inventory accumulation figure of, say, $6 billion at annual rate. This means that if inventory accumulation continued at the then current pace for four quarters or a full year, then the total addition over this 12-month period would amount to $6 billion. Or put another way, a $6-billion accumulation at annual rate implies that the actual accumulation over the quarter in question came to about $1½ billion.

One-shot Factors

Not everything that shows up in a business statistic such as sales or production can be attributed to basic or underlying economic forces. Weather, strikes or threats of strikes, war scares, etc., can all play a significant role in fixing the magnitude of any given variable at any given time. And unless the analyst is aware of these interferences, there is always the possibility of drawing the wrong inferences from the reported results.

The problem becomes especially acute in the area of forecasting where, say, an unexpected strike can make for a sharp deviation in anticipated results. True, there is little that the researcher can do in advance to prepare for an unexpected happening such as a wildcat strike. Indeed, he may well claim that he is not responsible for forecast errors resulting from such events.

His argument is a good one—up to a point. The trouble, however, is that it ignores the fact that there is often ample warning of a possible strike—and this warning in itself can lead to predictable results.

Thus if wage negotiations are up for reopening in three or six months, the forecaster should be expected to make allowances for heavy inventory hedging before the deadline and equally heavy inventory paring after the deadline has passed.

When this type of situation prevails, the analyst is usually on pretty firm ground, whether there is a strike or not. The hedge buying will take place in any case. Similarly, the falloff in demand once the contract deadline is passed is inevitable—even if a strike can be avoided.

Weather is perhaps even a less predictable hidden influence. A shift of a few degrees in summer can make a big difference in an item such as beverages where demand closely follows the rise and fall of the mercury. Similarly, the success of a spring or fall clothing line may well depend on whether or not there is normal, seasonal weather.

Moreover, the weather factor needn't be limited to consumer lines. In the construction industry, for example, rainfall and temperature usually determine how much cement, brick, glass, and lumber will be sold in any given month. And in a more general area, severe winter weather has been known to play havoc with production in certain areas of the country.

International political events add still another dimension to the real meaning behind a reported figure. Fear of devaluation can send people into precious metals as a hedge—with the price of silver, gold, and platinum shooting up to points far above those justified on the basis of normal supply and demand market levels.

War scares present a similar forecasting problem. The bulge in big-ticket consumer sales in 1950 and 1951 was not due to any sudden surge of consumer opulence, but rather to scare buying engendered by the Korean War. These same war-fear-connected factors also result in price run-ups on commodity exchanges—spurts which are usually followed by price declines of equal magnitude when fears of a bigger crisis tend to die down.

Then, too, a political influence on statistical data need not be connected to war. Thus in 1969 Red China suddenly stopped selling antimony—a key metal used in plastics, in paints, and as a flame retardant. Since China normally supplied half of the world's needs, a shortage developed and prices skyrocketed. Between early 1969 and early 1970 tags nearly quadrupled—knocking cost estimates of a lot of antimony users into a cocked hat.

Political agreements or understandings can also affect data. The sharp drop in United States steel imports during 1969 was directly attributable to a voluntary quota agreement on the part of foreign sellers—and not to any diminution in demand on the part of United States buyers. Similar pacts have made for sharp shifts in the importation of certain textile products.

The receipt of one big order can also sometimes play havoc with the analyst trying to plot the underlying business trend for a firm. Take the typical problem of a firm's largest customer, who always tends to place his order at the same time of the year. This will invariably distort the monthly pattern of incoming orders—though of course it has little meaning as far as signaling a basic change in the company's long-run performance is concerned.

But whatever the hidden effect and whether the analyst can or cannot do anything about it, knowledge of what is behind these figures is a must. Thus if sales for some noneconomic reason jumped 15 percent over the past quarter, cognizance of this fact would prevent extrapolation of this bulge into the future.

While obviously no two situations are exactly alike, some basic strategies in dealing with occurrences like these are called for. The following are among the most popular.

1. The elimination approach. Using this approach, a year of severe drought, for example, might be lifted out of a time series in order to calculate more accurately the long-term trend in the production of a basic food item such as wheat.

2. The adjustment approach. This involves raising or lowering the reported figure to bring it into line with what might have occurred in the absence of noneconomic interference. Thus steel production might be adjusted downward during a hedging period and upward during the subsequent stock-paring period. As a result, the adjusted figures would reflect normal operating conditions.

3. The moving-average approach. By averaging each reported result with those occurring immediately before and after, any disturbing noneconomic effect tends to be diluted. This approach is generally used to smooth out the "one big order" type of distortion referred to above.

4. The range approach. Where weather can influence performance, a range of projections can be given. Each one would represent a different set of assumptions regarding temperature and precipitation. Each such projection can also be accompanied with probability estimates based on past experience.

5. The asterisk approach. Here nothing is done to the reported figure. Instead, an asterisk or other identifying symbol is attached to the number to alert the reader that special one-shot influences were operating at that particular time.

Failure to use one of the above approaches can lead to faulty comparisons with other periods—and possibly a serious miscalculation of the underlying movement or trend. A more detailed discussion on some of the comparison pitfalls stemming from such one-shot events may be found in Chapter 13.

Productivity Effects

Efficiency or lack of efficiency often has a major effect on business data. And unless these productivity trends are taken into account, the analyst may well be led into making an erroneous conclusion.

Take the case of services and their role in the United States economy over the past few decades. A cursory glance at the jump in the percentage of total United States manpower devoted to services would seem to suggest the rising importance of this key economic area.

But closer examination would indicate that this development is due primarily to the slow rate of growth in service productivity, compared with that in industrial productivity—a fact which has forced the use of a greater number of workers in the service area in order to achieve a comparable gain. Indeed, this is verified by the service share of real United States output. This has remained virtually unchanged over the period under study. All this is summarized in Figure 11-4.

Incidentally, because of low productivity, service prices have also tended to rise substantially faster than product prices. Thus dollar service outlays would tend to make up an increasing share of the economy's overall outlays—again a possibly misleading figure if one were to use it to gauge the relative importance of services in our economy. And this is why Figure 11-4 compares the service share of output in real, or constant-dollar, terms.

Productivity also tends to influence some statistics by its relatively pronounced cyclical behavior. Historically, when production begins to tail off, productivity gains follow suit. And in times of a severe downturn in business, a negative productivity factor is not uncommon.

There's a good, solid explanation for this. When output slows down, there is a lag in the labor-force response. Several months may elapse before workers are actually dropped from the payroll. Equally important, production rigidities also generally rule out a percent reduction in the labor force commensurate with a percent reduction in production, no matter how much time is allowed to elapse. Ergo, the ratio of workers to production tends to rise, leading to the inevitable decline in output per man-hour.

Then, too, overhead workers are needed whether 100 or 1,000 units are turned out per week. In other words, some labor costs are necessarily fixed rather than variable. Again the inevitable conclusion is that the greater the production slowdown, the greater the downward pull on productivity.

Finally, note that when labor is scarce, manufacturers may elect to hoard skilled workers rather than let them go. They thus avoid the

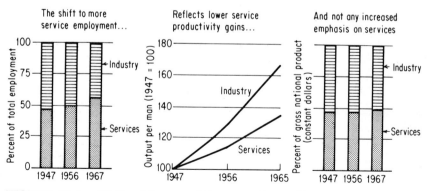

FIGURE 11-4   (*National Bureau of Economic Research*).

problem of having to recruit and retrain them some three to six months later, when business is likely to bounce back. This is precisely what happened in early 1969 following the Vietnam labor shortage and boom. Specifically, production and sales leveled out, but employment continued to creep up. Indeed, unemployment actually fell during the early part of this period. This seeming paradox was later explained by the reluctance of management to cut down on its work force in what was then deemed to be a chronically tight labor market.

For the statistician, all the above means that any near-term projection of manpower requirements must be tied not only to changes in production schedules but also to likely changes in productivity.

Individual firm manpower planning must also contend with still other productivity variations—ones due to start-up difficulties. Thus if a new automated factory is expected to beef up efficiency by, say, 15 percent, this goal may not be actually achieved for upward of a year after the plant comes on stream—or until the usual production bugs have been ironed out.

Cost estimates, too, are affected by temporary productivity shifts. That's because unit labor costs are closely related to productivity. Other things being equal, the bigger the productivity gain, the smaller the amount of upward pressure on unit labor costs. Reduced efficiency gains (which usually occur during periods of business slowdown), on the other hand, mean increasing cost pressures.

This, in turn, suggests planning on somewhat higher-than-normal unit labor costs during periods of sluggish activity. On the other hand, there's little need to panic when these costs do indeed rise during such periods—for this is probably due to cyclical factors rather than any deterioration in the firm's underlying long-run efficiency trend.

TWELVE

Percentages:
The Do's and Don't's

Percentages provide the key underpinning for much of today's statistical analysis. Index numbers, interest rates, profit margins, seller's markups, discounts, rates of growth—all these and many more of today's commonly accepted yardsticks are simply variants of a percentage, for all depend on the simple idea of comparing the magnitude of one number with that of another.

In some cases the percentage technique is a virtual must. Take the case of evaluating returns on two investment alternatives. Investment A may well bring in more dollars than investment B—but only because investment A is much larger. The correct approach in this instance, of course, is to compare the percent return on investment A with that on investment B.

Indeed, wherever size differentials are involved, the percent approach is the only meaningful one. Thus the comparison of sales of an individual company with those of an industry can be accomplished only by laying their percent changes side by side. This is another way of saying that rates of change rather than absolute changes are often crucial elements in statistical analysis. This point was also made in Chapter 3, where ratio charts (stressing percent change) were compared with arithmetic charts (stressing unit change).

Percentages are also essential simplifiers, reducing large numbers into smaller figures that the mind can grasp more easily. Indeed, in many ways the percentage approach is often little more than just good old plain common sense. Nevertheless, difficulties do arise—primarily because percentage notation can be used to distort, hide, or perhaps only give one side of a many-faceted operation. It is to these types of problems that much of the present chapter is addressed.

Some of the difficulties and pitfalls have been alluded to elsewhere in this book, though generally in abbreviated form. Below, however, is a more comprehensive discussion—aimed at putting all the relevant caveats into proper perspective.

Percents and a Changing Base

Perhaps the most common misconception involving percentages is that a given decline and a subsequent advance will put you back at the starting point. A moment's thought will illustrate why this cannot be so. Assume a 50 percent decline from 100—followed by a subsequent 50 percent advance. The decline would bring the figure down to 50 (50 percent of 100). The subsequent advance would bring the number up to only 75 (50 percent of 50 + 50). Indeed, to get back to the original level, the increase would have to be 100 percent—double the original decline.

The problem here is basically one of a changing base. The 50 percent decline applies to a base of 100. The 50 percent advance applies to a base of only 50. Ergo, the end result is a lower final figure. Based on the foregoing, it should also be clear that the magnitude of possible discrepancy is directly proportional to the size of the percentage involved. A 1 percent upward and downward movement will put one back close to the original level because the base has been changed only slightly.

A much more serious distortion arises in the conscious choice of one base over another one. The choice is often dictated by the impression the analyst wishes to leave. The most common example of this is provided by the retailer who quotes markup as a percent of the sales dollar rather than as a percent of cost. Since the sales-dollar base is larger than the cost base, the markup seems smaller—which gives the impression that the retailer is making a smaller profit. This practice has become so entrenched in the trade that a retail margin is now automatically assumed to stand for markup per dollar of sales.

Yet the profit per dollar of cost is clearly a more meaningful yardstick, for it gives the return on investment—the basic measure of any firm's overall financial health.

A somewhat similar approach is used by manufacturers who prefer to quote profits as a percent of sales rather than as a percent of investment or stockholders' equity. Again the reasoning is the same: Margins are almost always smaller when using the sales gauge—and hence are preferred by firms wishing to establish the need for still higher margins.

A look at recent government statistics will verify this relationship. In 1969, manufacturing after-tax profits per dollar of sales (based on SEC-FTC data) came to only 5.1 cents. The comparable after-tax earning figure expressed as a percent of stockholder's equity came to more than double this figure—about 12.1 cents.

This is in no way meant to imply that profits per dollar of sales is not a useful concept for some purposes. Certainly, a look at both types of margin can provide helpful hints on such interrelated profit determinants as capital requirements and sales turnover rates. Nevertheless, it is equally true that when margins are used to establish profitability, it is earnings per dollar of investment that is the more meaningful yardstick.

International currency revaluation is still another area where base switching is used to point up what the analyst wants to emphasize. The 1969 global wave of currency changes can best illustrate this point. When Germany increased the value of the mark in October 1969, that country's officials took the number of marks to the dollar under the new rate (3.66 marks to the dollar) and compared it with the previous rate (4 marks to the dollar). They then computed the percent decline (0.34 marks divided by 4 marks) and came up with an 8.5 percent upward revaluation. In other words, they were saying that the dollar (after revaluation) would buy 8.5 percent fewer marks.

Meanwhile, the International Monetary Fund figured the other way around—calculating in terms of what the mark would buy. With the original parity of 25 cents (4 marks to the dollar) rising to 27.3224 cents (3.66 marks to the dollar), the IMF figured the percent upward revaluation at 9.3 percent (27.3244 divided by 25). In other words, the mark (after revaluation) bought 9.3 percent more dollars.

Why did the Germans choose to emphasize the first percentage? The answer is clear. There was considerable political and business opposition to revaluation in Germany at that time, and the authorities wished to play down the size of any revaluation.

A few months previous to that, a French devaluation resulted in a similar choice. The devaluation, in terms of the number of additional francs a dollar would buy, was put at 12.5 percent. But in terms of the fewer dollars a franc would buy, this dropped to nearly 11.1 percent. The French chose to emphasize the former to show how much they were doing and sacrificing to correct their then serious balance of payments deficit.

Sometimes a base is changed not because one wants to but only because it is extremely difficult to avoid doing so. Thus one trade publication widely touts the fact that it can provide its readers with percent changes in sales in its specific field on an up-to-date basis. What the publication fails to tell its readers is that its survey is not comparable on a month-to-month basis because of a changing base.

Specifically, what the magazine does is to query its readers every month on total sales during that month, during the previous month, and a year ago; then it computes the percent change from month- and year-ago levels. Unfortunately, not everybody replies, and those who reply one month may not do so the next. Thus this magazine cannot possibly build up an index tracing the level and percent changes in sales from month to month over an extended period of time.

About the only use the survey has is to give a "quick and dirty" picture of the situation for the month in question. It is next to useless as far as pointing up changes in the underlying trend over any meaningful period of time.

Perhaps the most common type of base-changing distortion involves the shifting of a base year to "prove" a point. Thus a labor union trying to show a sharp gain in management profits will compare current earnings with those realized during the last recession year. Similarly, management will choose its best previous showing to emphasize that its profit gain over the interim period has been negligible.

Politicians are also past masters at this ploy. The party out of power will invariably choose a base to show a sharp rise in prices—and hence will establish its argument that the incumbent party is responsible for inflation. The party in power, on the other hand, usually takes the opposite tack—choosing the base that can best illustrate stability. Or if that's impossible, they will choose a base that shows relatively modest inflation and compare it with a runaway inflation period when the opposition was last in power. Similar approaches can be used to prove one's point on unemployment, purchasing power, or any other economic measure which has meaning to the average voter.

Alternative Ways of Looking at Change

Another method for emphasizing what is deemed to need emphasizing involves the citing of several different versions of a current change—and then comparing all of them with a fixed base. The technique results in the presentation of several alternative percentages—and permits the analyst to then stress what he regards as the real or true measure of change. This can probably best be illustrated by current auto industry marketing practices.

Virtually every fall, when auto makers introduce new car models, they are faced with the problem of making a price increase palatable to both buyers and government price watchers. Thus they attempt to present their increase in a way which will seemingly yield the smallest percentage advance. A General Motors release accompanying the introduction of 1970-model cars (see Table 12-1) shows how it's done.

Note first that the release contains three sets of calculations. The first and simplest method is simply to take the average 1970 price and compare it with the average 1969 price. Here this resulted in about a $119 average increase—the equivalent of a 3.9 percent price advance (see the extreme right of Table 12-1).

But then General Motors rightfully pointed out that their typical 1970 car was a different animal from their 1969 version—and hence the two were not strictly comparable. In the case of 1970 cars, the company stressed that these models contained better tires and other safety equipment which was deemed to be worth about $38 to the typical car buyer.

To make the cars comparable, reasoned GM, this $38 had to be subtracted from the price of 1970 cars or added to that of 1969 models. In either case, this suggests a smaller dollars-and-cents boost of $81. In percentage terms that's an increase of only 2.6 percent (see the middle column of Table 12-1).

TABLE 12-1   *1970 Auto Price Increases: Three Views*

			Composite avg. list price		
	Avg. equipped car	Avg. optional equipment	Base car with comparable standard equipment	Adjustment to 1969 models for changes made between standard and optional equipment	Base car
1970 model	$3,918	$729	$3,189	. . .	$3,189
1969 model	$3,838	$730	$3,108	$38	$3,070
Difference	$ 80		$ 81		$ 119
	2.1%		2.6%		3.9%

SOURCE: General Motors. For comparability, all the above amounts are weighted for estimated 1970 volume distribution by models and option usage.

Finally—and again with considerable validity—GM pointed out that the prices being quoted were for stripped-down models—models with no optional equipment such as radios, automatic transmission, etc. But the average buyer, said GM, normally buys a spate of these extras, which add up on the average to about $730. This sum would have had to be added to the dollar price of the car to have it reflect realistic buyer payout. However, since the price of these optional items remained relatively unchanged from 1969 to 1970, it then follows that the average percentage increase of a basic car plus extras should be something less than a mere basic car comparison would suggest.

With this in mind, GM came up with its third set of figures, which measure change from a base-price-plus-optional-equipment perspective on comparable models. As the first column of Table 12-1 indicates, the increase on this modified basis dwindles to only 2.1 percent.

As noted above, there is nothing basically dishonest in the above presentation. All it does is present the increase from three different vantage points—and then stresses the fact that the one with the lowest percent reading is the most realistic.

And stress this point GM did. Thus in the official press release GM started off by stating that it increased prices 2.1 percent—and then went on to explain how it reached this conclusion (see Table 12-1). Logic, on the other hand, would dictate starting at the 3.9 percent level and then working down in steps to the 2.1 percent reading.

Nevertheless, it should be pointed out that government statisticians subsequently backed up this Detroit appraisal. When Uncle Sam's official price index was posted a few months later, it showed only about a 2 percent advance. Bureau of Labor Statistics people at that time pointed out that car price indices—because they're designed to measure true price change—must be calculated on a fully equipped basis, adjusted for quality improvements. For a more detailed discussion of the role of quality in measuring price change, see Chapter 11.

It should also be noted in passing that the use of average prices to calculate change—no matter what vantage point one chooses—tends to be misleading. In the above case, for example, not all cars were advanced by the same percentages. At that time a certain segment of the public was on a cost-cutting kick—and was showing signs of switching to cheaper overseas makes. To capture this price-oriented market, GM kept prices on some of the low-cost 1970 models virtually unchanged. In effect, some people were getting their 1970 cars at virtually no increase, while others were paying upward of 5 percent more—facts that were hidden by the average percentages noted above.

But coming back again to different ways of viewing change, another often-used technique involves the "adjusting" of a current figure to take into account noneconomic factors—a point also alluded to in Chapter 11. If, for example, there were a major strike in a given month, then actual percentage decreases in, say, overall production might be reported—along with an alternative percentage showing what industrial production would probably have been if the strike had not distorted the figures. This is clearly justifiable—if one emphasizes that the adjusted figure is aimed only at showing the underlying trend.

In a similar vein, a large defense contract might conceivably result in a one-month sharp percentage spurt in new orders. To emphasize this fact, alternative percentages with the defense segment eliminated could better spotlight the underlying trend in the private economy. However, a more consistent method of dealing with such erratic order behavior might be through the use of moving averages.

Obscuring the Facts

Many times percentages are used as smoke screens to either hide the truth or give a misleading impression. A rundown on a few of the more popular of these ploys follows.

1. *The partial-information approach.* A typical example of this is the frequently made attempt to convince a reader or listener about the growing role of the average American as a stockholder. To give this impression, the analyst will stress the fact that about 30 percent of all American families own stock. This is certainly true. But what this statistic neglects to point out is that most people own only a few thousand dollars' worth of stock.

In short, the partial-information approach tends to give the misleading impression that 30 percent of United States families have a say in the running of United States industry. Obviously, this is far from true.

It would be much more meaningful in this case to present the above statistic along with a percentage of who owns how much. One such figure would show that the top 1 percent of families own about two-thirds of all privately held stock. Now the reader has enough information to reach an intelligent conclusion. He knows that while ownership is widespread (the message of the first, or 30 percent, figure), it is equally true that most holdings are small and that the large majority of the total is held by a small percentage of the population.

2. *The size of the base.* It doesn't take a mathematical wizard to realize that it is easier to increase, say, sales or profits from $1 to $2 than from $100 million to $200 million. Yet this fact is often glossed over when gains are presented in percentage form. Implication: Beware of rating a firm with a bigger percentage gain over a firm with a smaller one—unless you know the size of both firms involved.

Obvious? Sure. Yet invariably a wrong decision or conclusion will stem from the failure to investigate the size of the base. This often occurs in gauging international growth trends. A few years ago much ado was made of the fact that Russia was growing at a 10 percent annual rate while the United States was lagging far behind with only a 5 percent reading. Alarmist editorials appeared from coast to coast. What these writers failed to realize was that high percentage gains are almost always easier to come by when the base (in this case, the level of the GNP) is relatively small.

Even Russia has found this out. As her output has been increasing, that nation has been finding it harder and harder to maintain the same pace of growth. Thus by the late 1960s the highly advertised 10 percent Russian growth rate had slowed down to something near 7 percent. Again it wasn't that Russia had suddenly become less efficient. Rather, this figure reflected the facts of statistical life enumerated above—namely, that large percentage rates of growth become harder and harder to come by as the absolute figures grow.

It should be noted in passing that even while Russia was gaining at twice the United States rate, the absolute United States dollar gain was far bigger—again because of our bigger base. So in effect, the United States was more than maintaining her edge in the number of units of product turned out per year. In short, the absolute-dollar GNP gap between the two nations was actually growing at that time.

The fact that an absolute gap can be growing may sometimes be more important than the fact that a percentage gap is being narrowed. In such cases, looking solely at the percentage change may result in misleading conclusions. Many times this problem crops up when prices are involved. A simple example should help make this clearer.

Assume that Company A sells its product for $2 per pound, while Company B, because of higher productivity, can afford to sell its product for $1 per pound. Next assume that Company A's price rises 3 percent, while Company B's price rises 4 percent. At first blush it might seem that Company A is narrowing the price gap. But further analysis suggests otherwise if dollar prices are compared. Thus Company A (with a 3 percent increase) now sells its product for $2.06 per pound. Company B (with a bigger, 4 percent increase) now sells its product for $1.04 per pound. Before the round of price hikes Company B had a $1-per-pound advantage. After the increase it had a $1.02-per-pound advantage.

In short, the dollars-and-cents advantage—the key determinant of which company might get a given order—had actually widened, despite the fact that percentage trends suggested just the opposite. The different movements in absolute and percentage terms will occur only, of course, when the difference between the sizes of the bases is significant. In the illustration above, for example, one base was twice as large as the other—enough to permit a reversal of a seemingly narrowing price gap suggested by the use of percent data.

3. Official versus effective rates.   Often wide attention will be paid to announced percentages—even though it is known that such figures are unrealistic. Income taxes are a case in point. Much has been made of the fact that the percentage of income paid to Uncle Sam rises with the level of income—and that when income reaches high levels, the income tax rate can be in excess of 50 percent.

What is conveniently forgotten here, of course, is the fact that much of this large income is normally not subject to general tax rates—but probably comes under capital gains or some other tax provision which reduces the effective rate paid by the income recipient.

These avenues for reducing taxes are not confined to the rich, of course. But wealthy taxpayers do have more opportunity to manipulate their incomes to take advantage of special tax treatment. The resulting distortions tend to increase with income. Treasury studies estimate that, on the average, the amount of income open to special treatment goes up in about direct proportion to a taxpayer's increase in income, both because of the nature of tax laws and because of his efforts to make this so.

The surprisingly low effective rates at high income levels were highlighted a few years ago before the 1969 income tax reform law went into effect. According to the Bureau of Internal Revenue, under the old prereform law something around half of all taxpayers reporting adjusted gross incomes of more than $100,000 a year paid an effective rate of 30 percent or under. The basic or official rate at that time was 50 percent and over.

The concept of effective versus reported percentage is also often used in comparing percent earnings realized from two different types of income—each subject to a different tax rate. A good example is tax-exempt state and local bonds. Traditionally, interest rates are lower here. But from an effective perspective, they can be higher for wealthy individuals in high tax brackets. Thus if you are in the 67 percent tax bracket, a tax-exempt yield of 4 percent on a local (tax-exempt) bond is equal to a yield of 12 percent on a taxable security.

Such effective versus actual differences, of course, are by no means limited to the taxation area. Interest rates—particularly those involved in the so-called truth-in-lending law, discussed below—almost always present a problem of translating commonly quoted charges into effective rates. In the simplest case a $100 loan payable in a lump sum a year later with interest discounted in advance does not have an effective interest rate of 6 percent—but something more. Simple calculation, for example, would show that 6 percent for the use of $94 over a year is the equivalent of an effective rate of 6.4 percent—or as high as 11.5 percent if the "add-on" installment approach is used (see "Truth in Lending" below).

To a businessman, even a simple bank loan rate is not the truly effective rate for borrowing money—because of the necessity for keeping a certain amount of the loan, known as a *compensating balance,* on deposit at all times. For example, a loan agreement that provides for an 8 percent interest rate and a 20 percent compensating balance on an advance of $100,000 actually involves an effective rate of about 10 percent. Reason: The borrower is able to draw on only $80,000. In other words, he is paying $8,000 in interest for the effective use of $80,000—the equivalent of a 10 percent interest rate.

Because the effective cost of a bank loan is determined by the two factors—the stated interest rate and the size of the compensating balance—a bank can vary the effective rate on selected bank loans in the absence of a formal change in the prime rate.

4. *Hiding a small sample.* One often-used ploy—particularly in opinion research—is to quote the percent who are in favor or not in favor, without reporting the number of respondents involved. Thus a report indicating that 75 percent of the respondents favored brand A would be entirely unreliable from a statistical point of view if only four people had been queried.

Thus whenever percent figures are used, it is always a good rule to inquire about the size of the sample. Most reputable market researchers will normally include this number as standard operating procedure. Once this figure is known, it is also a good idea to inquire into the actual sampling technique—whether it was random, stratified, or something else. With these two additional bits of information it is then possible to assess the statistical reliability of the reported sample percentage.

5. *Exaggerating significance.* Some people will use a percent change to imply a significance which, in effect, may not exist. A report that profits declined, from, say, $1.6 million to $1.2 million may not ordinarily attract attention. But the same figure in percent or rate of return form—showing margins dropping from, say, 7.2 to 7 percent—is almost sure to be looked on as a significant development. It may well be significant. But then again it may not. In any case, the fact that it is in ratio form has no bearing on its importance.

The problem is basically psychological. There is generally thought to be something magic or learned about a percentage—something that implies statistical refinement or analysis. As such, all percentage changes are deemed automatically to be significant. Thus much ado is made monthly about a 0.1 percent increase in the cost-of-living or unemployment index—even though the range of statistical error is usually well above 0.1 percent.

Truth in Lending

Perhaps the most ambitious attempt to inject a note of honesty and realism into the use of percentages in the credit field occurred a few years ago when the federal government passed a comprehensive "truth-in-lending" law. The aim was to tell the borrower how much it was costing him—in terms of a true, or effective, annual interest rate—to borrow money.

Actually, in the world of big business it is standard operating procedure to state explicitly the annual interest rate charged. Business loans are always expressed this way. So are money market yields and home mortgage rates. Even the common savings account earns interest at a simple annual rate.

A problem crops up, however, when one enters the complex world of consumer finance. Take, for example, the common practice of making a periodic charge on credit cards or revolving credit accounts. Before passage of the truth-in-lending law, the charge was invariably reported as, say, $1\frac{1}{2}$ percent per month on the unpaid balance. Conspicuous by its absence was the obvious statement that such a charge is the equivalent of an 18 percent annual interest charge.

To be sure, most users of revolving credit don't let their bills remain unpaid for such a long period of time. Nevertheless, it may be that when a consumer knows he is paying at an 18 percent annual rate, he may well seek to borrow the money more cheaply elsewhere and use the proceeds to pay his charge account within the monthly time limit, before interest is charged.

To facilitate this kind of rational decision, the new law requires that the annual interest charge be clearly stipulated when a loan is made. In general, when computing such a true interest rate, creditors must use the so-called actuarial method—the method used in making mortgage loans. The actuarial method involves a redistribution of the dollar finance charge in accordance with a series of declining unpaid balances under an installment contract. It specifically outlaws, then, the use of the add-on or discount percentage to denote interest cost.

These latter two methods are often misunderstood by unsophisticated borrowers—leading them to believe they are getting money a lot more cheaply than they really are. In short, both these methods grossly understate the true rate.

Both of these misleading methods can be illustrated in terms of a $100 bank loan, with the bank quoting a seemingly low 6 percent rate. Under the add-on method, the $100 is received, and the loan is repaid in 12 equal monthly installments that add up to $106 over the year. Hence the bank may claim that the borrower is paying only 6 percent more than he borrowed.

But note that the borrower has the effective use, on the average, of only about half the $100 over the 12-month period. (Remember he is repaying part of the principal with each monthly installment.) A more realistic calculation based on the actuarial method would reveal that the true, or effective, annual interest rate is 10.9 percent—nearly double the seemingly small 6 percent rate quoted by the bank.

The effective annual rate is even higher under the alternative discount method. Specifically, the borrower receives only $94 when asking for a $100 loan. The bank says it is discounting the loan—taking its 6 percent, or $6, interest charge out in advance. The borrower is then expected to pay back the $100 principal in 12 equal monthly installments.

Again the borrower has the effective use, on the average, of only half the borrowed amount over the year. But this time he is paying $6 for the privilege of getting his hands on only $94. Contrast this to the add-on method, where he is paying $6 for the privilege of getting $100. Ergo, the discount method carries a somewhat higher interest charge—11.5 percent, compared with the 10.9 percent under the add-on method.

All this, however, has ended with the passage of the truth-in-lending law, which requires that the effective annual rate—either 10.9 percent or 11.5 percent—be quoted to the borrower at the time the loan is made.

Complications, of course, arise where repayments aren't equal. To simplify matters, special percentage rate tables have been prepared by the Federal Reserve Board and others to meet such special needs. The Federal Reserve version covers 200 pages.

Meanwhile for those who want a "quick and dirty" way of calculating rates on loans such as those discussed above, mathematicians have derived this simple formula:

$$R = \frac{2MI}{P(N + 1)}$$

where R = effective interest rate
  M = number of repayment periods per year
  I = interest charge in dollars over the life of the loan
  P = original amount borrowed
  N = number of payment periods over the life of the loan

Substitute in the add-on bank loan illustration given above, and the following results are obtained:

$$R = \frac{2\,(12 \times 6)}{100(12 + 1)} = \frac{144}{1300} = 10.8\%$$

To be sure, this is not as precise as the exact 10.9 percent reading yielded by strict adherence to actuarial techniques. Nevertheless, it can be seen that the calculations are relatively easy—and that the degree of variance is relatively small.

To many, these effective annual bank loan rates may seem high—and they are. But in all fairness it should be pointed out that banks in many cases are still your best bet for consumer-type loans. Their charges certainly run far below those quoted by the typical finance company. On the other hand, there are two limited sources for cheaper borrowing: (1) passbook loans, which involve borrowing against money on deposit in a savings bank, and (2) life insurance loans, which can often be obtained at a straight 5 percent interest rate—far below the prevailing market cost.

But to come back again to truth in lending—the new intelligence, in effect, has given borrowers several additional options. Instead of snapping at the first loan offer because of either ignorance or lethargy, borrowers may now (1) elect to wait for rates to come down a bit before making a commitment, (2) shop around carefully for the cheapest loan, (3) decide to make a larger down payment, or (4) conclude that it is cheaper to draw the needed money out of the bank and thus avoid these sky-high interest rates.

If nothing else, the new truth-in-lending approach makes potential borrowers stop and think before plunging ahead. To be sure, a new refrigerator ticketed for $300 may appear to be a good buy when the dealer quotes a low monthly amount to be repaid over a period of a year or so. But it might seem a lot less acceptable when the potential buyer realizes the huge finance charges that he is being asked to assume.

Percentage Change versus Percentage-point Change

Movements of indices from one date to another are usually expressed as percentage changes rather than as changes in index points, primarily because index-point changes are affected by the level of the index in relation to the base period, while percentage changes are not. To be sure, when the index-number magnitude is near 100, the difference is negligible. Note, for example, that an index-point rise from 101 to 102 (1 point) is only fractionally above the true percentage increase.

Problems arise, however, when the level is far above or below the 100 level. The following example, culled from a recent Bureau of Labor Statistics release, illustrates the two different computations involved in calculating the index-point and the percentage changes:

Index-point change		*Percentage change*
September 1969 CPI (1957–1959 = 100)	129.3	Index-point difference divided by
Less August 1969 index	128.7	the index for the previous period:
Index-point difference =	0.6	$\dfrac{129.3 - 128.7}{128.7} \times 100 = 0.5$ percent

Note, too, that the difference tends to grow as the magnitude of the index number grows. Thus an index-point rise from 200 to 202 (2 points) is the equivalent of only a 1 percent rise. Similarly, an index-point rise from 400 to 404 (4 points) is again the equivalent of only a 1 percent advance.

This is one reason why statisticians prefer periodic updates of index-number base periods. Shifting the base to a more current period brings the actual index readings back closer to 100—and thus reduces the possibility of misunderstandings between percent and index-point changes. Such an updating also reduces the temptation to distort. Thus if the index reading were well above 100 during a period of inflation, people who might want to play up the price spiral would be tempted to quote the percentage-point advance—conveniently forgetting to give (1) the true percent change or (2) a base-period reference, which would allow the reader to compute such a true percent change on his own.

Percentage points can be used with somewhat more validity in opinion research areas. It is perfectly legitimate to talk, say, about a 4-point spread between the 52 percent who prefer brand A and the 48 percent who opt for brand B. But even here some caution is advisable. For example, it should always be kept in mind that a 2 percent preference

swing can completely wipe out this 4-point spread. The difficulties encountered by political pollsters can attest to this kind of percentage-swing
problem.

It should be noted in passing that not all this uncertainty in opinion
research stems from the fact that people change their minds. A good
part may well be due to the fact that the approach depends on sampling
techniques. As noted in Chapter 5, chance variation occurs in all samples. But this latter type of variability can be adequately accounted
for by using statistical probability. Thus, given the sample details, it
may be possible to state that 9 chances out of 10 the majority of people
prefer brand A when a sample reading of 52 percent is obtained.

In any case, it is important to separate out the variability due to
statistical chance and that which may come about as a result of subsequent change in opinion. The former is predictable, and the latter is
not.

Marginal versus Average

In some instances percentages can be expressed in marginal or average
terms. In the area of profits, for example, if after-tax earnings are $8
million and sales are $100 million, we could say that average profits
were 8 percent of sales. But this concept is of limited value in decision
making. Most businessmen have to make marginal decisions—say, to
accept or reject an additional sale. In this latter kind of problem it
is the additional profit relative to the additional sale that is the pertinent
point.

Thus if the rate of profit on the additional order were to turn out
to be 9 percent, the businessman would probably be more than eager
to accept it. On the other hand, if the rate of profit on the additional
order were only 2 percent (because of the need to put men on overtime,
use less efficient machines, etc.), then he might think twice before accepting it. In short, the businessman cannot automatically assume the same
profit performance on the margin that he has experienced on the
average.

This might seem obvious. But that's only because the example is obvious. In many cases everyday business decisions are made on the basis
of average performance—with the implicit assumption that the average
will continue to apply to future moves. This is not always warranted.

The difference between average and marginal can best be illustrated by looking at overall consumer buying behavior. It can be shown, for example, that consumers spend, on the average, about 94 to 95 percent of their disposable income. Hence the average "propensity to consume" is said to be 94 to 95 percent. On the other hand, it can also be shown that each dollar of additional income received by the typical consumer will result in about 80 cents in additional consumption. Hence the marginal propensity to consume is said to be about 80 percent.

For those who are mathematically oriented, the marginal concept can most easily be likened to the slope or b value in the straight-line equation $Y = a + bX$. The b value—since it measures the change in Y per unit change in X—is a marginal measure.

The use of marginal analysis, of course, need not be limited to percentages. Marginal sales, costs, and profits can be expressed equally well in terms of absolute numbers. In either case, the use of the marginal concept is recommended whenever evaluation of change is involved.

The Compounding Effect

Since percentages—particularly interest or growth rates—are generally small numbers (10 percent or lower), there is sometimes a tendency to shrug off small differences as inconsequential. What is forgotten, of course, is that the process of compounding (applying a growth or interest rate in a given period not only to the original figure but also to all the increments of preceding periods) can cause a small percent difference in a rate to balloon into a yawning gap over an extended period of time.

EXAMPLE: While annual GNP growth rates are surprisingly close for most industrial countries, a fraction of a difference per year can make for an enormous change in living standards over just a few generations. If one begins with a level of 100 for example, a compound growth rate of 1½ percent per year over a century would yield a figure of 441; a growth rate of 2 percent, a figure of 725; and a growth rate of 3 percent, a figure of 1,922.

To see how much of these big differences is due to the compounding effect, calculate the outcome at the end of a century on the basis of a simple growth rate (multiplying the annual rate by the 100 years involved). The figures would be 250, 300, and 400, respectively. That's in sharp contrast to the much larger 441-to-1,922 range obtained by going the compound route.

Note, too, that the compound effect tends to grow with the size of the rate involved. At a 1½ percent rate, the compound figure is almost twice as large as the simple one. At a 3 percent rate, the compound figure is nearly five times as large.

To put the compounding effect into proper perspective, some people often rely on this rule: To determine the number of years required for anything to double, divide the annual rate of increase into the number 70. Thus if the economy were growing at a 10 percent annual rate, it would double in approximately 7 years (70 ÷ 10).

Much of the insidious effect of creeping inflation can be traced to the compounding phenomenon. Thus when the price level is rising 5 percent a year, it may seem only moderately annoying—but certainly not catastrophic. Yet look at what would happen if this trend continued for 10 years. After a decade of such constant price uptrend, the general price level would have zoomed up by a hefty 63 percent.

The same effect can be expressed in terms of lost purchasing power. Table 12-2 shows how long it would take for the dollar to lose half of its purchasing power at differing small rates of inflation. At the 5 percent rate, for example, half of the buying power of a dollar bill would have been wiped out in about 14 years.

TABLE 12-2 *Half-Life of the Dollar under Inflation*

Annual rate of price increase, %	Years needed to halve the dollar's purchasing power
1	70
2	35
3	23
4	18
5	14

A similar situation exists where mortgages are concerned. In a sense the mortgagee often doesn't know the sum total of obligations he is assuming. True, the lender will tell him his true annual rate of interest. But left unmentioned are such key facts as the cost of the mortgage over the life of the loan. Thus if you were to assume a 30-year $20,000 mortgage at a 10 percent interest rate, you would be paying out $43,186 in interest over this period. In other words, you would be paying back $63,186 on a $20,000 loan.

This amount could be reduced, of course, by reducing the maturity of the mortgage and/or reducing the interest rate. Since interest rates are usually given, maturity reduction is the only way out left for most people. Thus reducing the above $20,000 mortgage (10 percent rate) from 30 years down to 20 years would result in an interest charge totaling only $26,322 over the shorter period—well under the 30-year figure of $43,186.

As for the effect of differing interest rates on the total dollar interest cost, you can see the results in Table 12-3. It shows varying dollar interest payments on a $10,000 25-year mortgage. Note how a small change in the interest rate again can make for a major difference in the overall dollar interest payment.

TABLE 12-3 *Interest Payments on a $10,000, 25-year Mortgage*

Interest rate, %	Interest payment
8	$13,154
9	$15,176
9½	$16,211
10	$17,261
10½	$18,325
11	$19,403
11½	$20,404

CONCLUSION: Other things being equal, a home buyer should aim at (1) the lowest interest rate possible, (2) the lowest feasible maturity, and (3) the biggest feasible down payment—for in today's period of relatively high mortgage rates, the less you borrow, the better off you are. It may also be worthwhile to get the privilege of prepaying mortgages without penalty (to cut down on costs in the event of a personal cash windfall or significant decline in mortgage rates). Finally, avoid mortgage financing of short-term durables such as appliances which have a life of only 5 to 10 years.

Percentages and Leverage

Anyone looking for a quick buck is more often than not looking for an investment with substantial financial leverage. The term "leverage" in this context has precisely the same connotation it has in the physical sciences, where a small input of work at one end of the spectrum results in a large output at the other end—with the ratio of input to output defined as the degree of leverage.

Carrying this analogy through to the financial sphere—if one can obtain ownership of an asset with (1) only 10 percent cash and (2) a loan from the bank for the rest, then the asset holder is said to have financial leverage. He is making his dollar go a lot further than one who purchases an asset outright. This, of course, is the raison d'être of margin purchases in the stock market. Indeed, 40 years ago (before the 1929 crash) as little as $10 down brought buyer control of $100 or more of stock.

To be sure, this has all changed. Today, for example, a minimum of nearly 80 percent is required in cash payment for stock market securities. On the other hand, margin requirements are still very low in the commodity futures markets. Hence, there is still plenty of opportunity for making a lot with a little through this kind of leverage technique.

In today's futures markets, for example, one has to put up only 5 to 15 percent of the value of a contract, depending on the particular commodity involved. But because buyers normally unload their original commitments before the delivery date, speculators rarely have to ante up the full cash value of their contracts.

At the same time they get the full market play during the period they hold a contract. Thus in a commodity where the margin requirement is only 5 percent, a 5 percent price rise will double the speculator's money. A simple example can illustrate this point. Assume that commodity X sells for $1 a pound and that the investor, through the use of futures, buys 1,000 pounds with a 5 percent down payment of $50. Next assume that the price rises to $1.05 a pound. He sells, making a 5-cent-per-pound profit on the 1,000 pounds which he owns—for a total profit of $50. In short, a $50 original investment has resulted in a $50 profit. Conversely, a 5 percent price drop will wipe out his original investment. "There's no quicker way to make or lose a buck," is the way seasoned commodity traders usually put it.

The same opportunity for leverage—a chance for big profits with only a small investment—holds for stock market "puts" and "calls." These are options to buy or sell stock which are in many ways similar to dealing in a commodity futures market. Assume, for example, that a speculator feels the stock of the ABC Corp. will go up—but he doesn't want to tie up a large sum of money by purchasing the stock outright.

Instead, he decides to buy a "call option." This gives him the right to buy 100 shares of the stock during a specified length of time at an agreed-upon "striking price"—usually the market price at the time of the option agreement.

To obtain the option, the buyer pays a fee or premium, generally ranging between 5 and 20 percent of the current value of the stock. If during the option-to-buy period the price does indeed go up, the holder of the call option can then buy 100 shares at the price stipulated in the option agreement and sell them on the open market at a higher price—thereby making a quick profit.

In any case, he has invested only 5 to 20 percent of the value of the stock and has been able to reap the same profit (less the premium) that he would have achieved had he purchased 100 shares outright. Of course, if the price drops, he doesn't exercise his option right to purchase the 100 shares. Then he's out the 5 to 20 percent premium —but that's the maximum limit of his loss.

Purchase warrants to buy stocks provide an even more dramatic example of financial leverage. The price of this commercially traded paper tends to fluctuate by a much greater percentage than the price of the stock that it can buy. Ergo, there is the possibility of a big killing—or big loss if you're unlucky.

An example again can best illustrate this type of leverage. Company A, whose stock is selling for $10 a share, issues 10-year warranties with an "exercise price" of $12—entitling the holder to buy the stock at $12 anytime over, say, the next 10 years. Next assume that the stock rises to $15. The warrant also rises, say, from an original zero value to $5. Some $3 of this represents the immediate profit a warrant holder can make by trading it in for $12 worth of stock and then selling it back on the market for $15. The other $2 represents a premium paid on the possibility that the stock will go even higher before the warrant expires.

Assume that a year later the stock price rises another $10 (to $25). The warrant follows suit, going up another $10 (to $15)—with $13 of this representing the profit if the holder exercises his option, and the other $2 representing the premium.

Note, however, what has happened in percentage terms during this last rise. While the stock and the warrant have gone up $10 each, the percentage rise in the warrant price is much greater because the warrant started from a lower base. Specifically, the warrant price went from $5 to $15—a jump of 200 percent. The stock-price advance—from

$15 to $25—represents a rise of only 66⅔ percent. This kind of leverage, of course, can work the other way, too. If stocks fall, the warrant falls more than the stock.

Examples of financial leverage and the risk involved need not be limited to the few given above. There are literally hundreds of different cases. But the point in all is essentially the same: A relatively small percentage change in one area (commodity futures prices in the first illustration and stock price in the warranty illustration) can bring about a big percentage change in the investor's rate of return.

Finally, it should be noted that leverage need not be limited to the investment arena. A small percentage reduction in costs, for example, can be said to have much more leverage on profits than a similar percentage increase in sales. That's because most of the added sales dollars are offset by higher production and distribution costs. Indeed, this is why cost cutting has become such a key plank in recent corporate drives to beef up profits.

Such cost cutting can cover any number of different operations—including purchasing, production, distribution, and transportation. In the case of purchasing the leverage can be surprisingly high—primarily because purchasing accounts for such a large percentage of the sales dollar.

Table 12-4 vividly illustrates this point. Thus if your firm devotes 50 to 54 percent of its sales dollar to procurement and its normal profit is 6 percent, then it can be shown that a 2 percent savings in procurement is the equivalent of a hefty 17½ percent increase in sales. It goes without saying, of course, that the higher the percentage of sales devoted to procurement and the lower the normal profit ratio, the greater the leverage.

Percentages and Dynamic Change

Often a percentage which might be significant during one period of time may become relatively insignificant in a later period. Then again a percentage over an extended period of time may require a changing interpretation—despite the fact that the data are completely free of any errors and/or distortions. These problems of interpretation and evaluation usually occur because of the ever-changing structure of the business world.

Take unemployment. In the 1950s a 5 or 6 percent jobless rate was deemed to be normal. But not so anymore. Growing social pres-

TABLE 12-4 *Purchasing's Leverage on Profits*

(Sales increase needed to equal a 2 percent procurement cost savings)

Company's net profit	Percent of sales spent on material procurement									
	35–39	40–44	45–49	50–54	55–59	60–64	65–69	70–74	75–79	80–84
5%	15.0%	17.0%	19.0%	21.0%	23.0%	25.0%	27.0%	29.0%	31.0%	33.0%
6%	12.5%	14.2%	15.8%	17.5%	19.2%	20.8%	22.5%	24.2%	25.8%	27.5%
7%	10.7%	12.1%	13.6%	15.0%	16.4%	17.9%	19.3%	20.7%	22.1%	23.6%
8%	9.4%	10.6%	11.9%	13.1%	14.4%	15.6%	16.9%	18.1%	19.4%	20.6%
9%	8.3%	9.4%	10.6%	11.7%	12.8%	13.9%	15.0%	16.1%	17.2%	18.3%
10%	7.5%	8.5%	9.5%	10.5%	11.5%	12.5%	13.5%	14.5%	15.5%	16.5%
11%	6.8%	7.7%	8.6%	9.5%	11.5%	11.4%	13.5%	14.5%	14.1%	15.0%
12%	6.3%	7.1%	7.9%	8.8%	10.5%	10.4%	12.3%	12.1%	12.9%	13.8%
13%	5.8%	6.5%	7.3%	8.1%	8.8%	9.6%	10.4%	11.2%	11.9%	12.7%
14%	5.4%	6.1%	6.8%	7.5%	8.2%	8.9%	9.6%	10.4%	11.1%	11.8%
15%	5.0%	5.7%	6.3%	7.0%	7.7%	8.3%	9.0%	9.7%	10.3%	11.0%

sures—plus increasing government recognition that everybody who wants to work should be able to find work—have changed all this. By 1970 anything significantly above a 4 percent rate was labeled "intolerable." In terms of evaluation this has resulted in a major shift. Businessmen 15 years ago could count on expansionary government policies only when the jobless rate went over the 5 to 6 percent range. These days, however, the alarm tends to go off anytime the unemployment rate rises much above the 4 percent level.

Also changing the meaning of jobless-rate statistics has been the increasing demand for decent, rewarding jobs with a promising future. Yet in figuring the unemployment rate, all jobs still count: part-time jobs, casual and intermittent jobs, dead-end jobs, menial jobs, and degrading jobs. This inability of the statistics to reflect the quality of jobs has in a sense limited their usefulness as a gauge of public policy. And other things being equal, it again makes a given level of unemployment less tolerable than heretofore.

Interpretation of factory operating rates has also changed substantially over the past decade. It used to be that whenever this yardstick of capacity utilization dropped below 87 percent, you could look for a sharp decline in plant and equipment spending. But again the ground rules have changed. Thus in 1969 capital outlays were upped by over 10 percent, despite the fact that factories were operating at only 82 percent of their potential.

Much of the change in investment attitudes can be traced to changing cost relationships. By 1969 the shortage of labor, plus its rapidly rising cost, had made automated cost-cutting equipment a lot more attractive—even where existing capacity was more than adequate.

Rapidly improving technology also played a role, making new investment almost mandatory in highly competitive industries. Newly developed mass markets are a third factor in speeding up the investment curve. Thus the sharp growth in computer and jet airliner outlays has been due, in part, to new mass applications which make it economical to buy substantial quantities of newer, larger models, even though the ones in use are still quite new and efficient.

Sometimes capacity rates take on a different significance because of changes in the demand mix. Thus a few years back there developed what appeared to be a big bulge in polyvinyl chloride (PVC) plastic capacity. But this proved not to be the case; low prices for general-purpose PVC had moved producers to push specialty grades, which generally commanded a higher price. This trend reduced actual capacity because of short production runs and increased downtime, which were necessary to effect frequent product changeovers. Some industry observers said this chewed up about 10 to 20 percent of nameplate capacity. Put another way, a good deal of the glut was more apparent than real.

Price is still another area where old percentage relationships seem to be changing. The rate of annual price advance that a business firm can expect, for example, is now much greater than it was a few decades ago. Blame it on the greater accent on full employment growth and its resulting strains on demand and labor costs. Interest rates have also tended to seek a higher level over the longer pull—primarily because lenders now must recover some of their purchasing power lost by the eroded value of the dollar.

THIRTEEN

Comparability: A Matter of Apples and Oranges

What student hasn't been warned about the dangers of comparing apples and oranges? Yet many of today's most glaring statistical errors are of this sort: comparing the noncomparable. Much of this is due to the fact that this type of error can occur in so many unforeseen forms. For example, comparing pay rates in widget factory A with those in widget factory B may seem to be the most clear-cut of analytic procedures, and yet if factory A is automated and factory B is not, any such comparison is completely meaningless unless it takes into account the efficiency of each firm's work force. For it is unit labor costs, not wage rates, that ultimately determine which factory is better off.

Another example, culled from the equipment field, illustrates even more vividly how easy it is to fall into the apple-orange pitfall. A large machinery company, touting its influence, stated that it accounted for about 20 percent of all machine tool business. What its zealous public relations people had forgotten was the fact that one of its divisions (the largest) produced construction equipment rather than machine tools. It was only by adding its sales of both machine tools and construction equipment—and then comparing this total with the industry-wide machine tool sales figure—that the company came up with its 20 percent estimate.

When the error was discovered, the company recomputed the share-of-the-market ratio—taking its total of machine tools and construction machinery sales and comparing this with the comparable industry total of both categories. Computed this way, the firm's share of the market dropped to 9 percent. Further computations showed that if the company compared only its machine tool sales with industry-wide machine tool sales, the percentage also dropped—to 12 percent. Thus when the firm compared apples with apples or oranges with oranges, the figures weren't nearly as impressive as when it used the apple-orange comparison.

This chapter will focus on some of the more common forms of the many-faceted comparison problem. While it is virtually impossible to touch all bases, certain basic ground rules are applicable in most cases. Among the key do's and do'n'ts are the following.

1. *Make sure that what you are measuring is measurable.* Not everything can be forced into a quantitative straitjacket. And where it cannot, serious errors may result. Measuring the progress of the war in Vietnam by changes in "kill count" is a classic example of how this approach has led analysts astray. Put another way, avoid the simplistic approach of relating an easily measured statistic (e.g., kill ratio) with a basically complex problem (e.g., the war in Vietnam).

2. *Check to see that the statistics being compared measure the same phenomenon.* A comparison of the list price of product A and the discounted market price of product B may well distort the actual difference. In any case, it's hardly likely to give the true relationship between the two items. Comparing an accounting method based on accelerated depreciation in one period with one based on straight-line depreciation in another can lead to similar problems.

3. *Compare on the same market level.* This is a variation of the "measuring the same phenomenon" problem. Comparing the wholesale price of eggs with the retail price of eggs is a typical example of this type of error. Comparing a price average of finished goods with, say, a price average of raw materials can lead to similarly questionable results.

4. *Watch out for statistical revisions.* Uncle Sam's statisticians

seem to take fiendish delight in updating old figures—usually to take into account new benchmark data. A current figure compared with unrevised back data may well yield different results from those which would be obtained by comparing that same current figure with revised back data.

5. Develop sound criteria for reporting if you are working up your own data. Well-thought-out sampling procedures and precise definitions of what it is to be measured are musts if a meaningful series of data is to be developed. The difficulty of measuring crime statistics over different periods of time (see the detailed discussion on page 267) is a case in point.

6. Be on the alert for one-shot distortions. A strike can distort a conclusion about how fast production is growing in a particular industry. Similarly, a severe winter storm can distort weekly trends in retail sales. Chapter 11 dealt with one aspect of this problem. A further discussion of how one-shot distortions can lead to faulty comparisons is presented on page 272.

7. Be sure a given reading today has the same meaning as a similar reading a few years ago. Changes in the product mix and changes in the way of doing business often affect the basic assessment of statistical results. Drawing on an example given in Chapter 12, a 6 percent level of unemployment was considered normal a decade ago but is thought to be intolerable today.

Sometimes unfavorable comparisons can be hidden by the "two-stage" ploy. For example, in late 1968, when auto companies wanted to play down price increases on new 1969 models, they announced one price advance at model introduction time—pointing up that that increase over the previous year was relatively small. Then a few months later they added a second increase to cover new safety equipment. They said the two-stage approach was necessary because the safety equipment was not ready for installation at the time of model introduction.

Whether this was true or not, the companies did manage to hold the initial increase to $50 to $65 per car—an amount that government price watchers found tolerable. Then by the time the second increase came around (some three months later), everybody had forgotten about the first one—and it went through with a minimum of adverse criticism.

Other ways in which the sophisticated auto companies can tone down price increases were discussed in Chapter 12.

The remainder of this chapter will examine these and other compari-

COMPARABILITY: A MATTER OF APPLES AND ORANGES **265**

son pitfalls in greater detail. But before leaving the general subject of comparability, it should be pointed out that statistical difficulties need not be limited to outright misinterpretation or deception. In some cases, comparison problems may simply be due to poor methods of presentation. Nonmathematically oriented executives, for example, often have difficulties in assessing change in statistical data expressed, say, in seasonally adjusted or constant-dollar or deflated form. The stumbling block is that the data may be presented in such an esoteric form that the nontechnical reader is unable to relate them to meaningful, action-oriented decisions.

Take the example of the constant-dollar price presentation, discussed more fully in Chapter 11. If selling prices are rising, the sales manager cares little that this "artificial" series shows lower figures than actual current-dollar sales results. To be sure, such executives are interested in why prices went up, but they are scarcely concerned about how many fewer dollars they would have received if prices hadn't risen. Similarly, if prices are falling, they are hardly interested in being told what the level of sales would have been if prices hadn't declined.

A similar problem occurs when artificial seasonally adjusted figures are introduced. On the other hand, as noted elsewhere in the book, seasonally adjusted and deflated yardsticks are necessary for intelligent decision making. How, then, can they be made palatable to businessmen? Here are some simple rules:

▪ When constant-dollar (deflated) sales figures are called for, convert the series into an index; then the deflated series can be labeled a "quantity" or "volume" index, which is a lot easier to grasp. This is certainly true when sales are involved because then the manager readily realizes he is being given a series depicting the number of units sold.

▪ When dealing with seasonally adjusted data, try to avoid showing them alongside the original data—at least in dollar form. This is confusing and seldom serves a useful analytic purpose. Rather, concentrate on seasonally adjusted figures if long-run trend analysis is your objective, and on unadjusted or raw data if financial statement analysis is your prime goal. Showing the seasonally adjusted data in index form might again be useful.

In short, a meaningful comparison can be a communications problem as well as a statistical one. Too much concern over the technical problems of manipulation and modification of data at the expense of clarity can easily be self-defeating.

Equating Shadow with Substance

Perhaps the most glaring misuse of statistical comparison involves attempts to measure the unmeasurable, quantify the unquantifiable. Both in business and in government, men of integrity, brains, and diligence have fallen into this trap. In most such cases self-deception rather than dishonesty is to blame.

The advertising man who measures a TV program's rating and then equates this with its ability to sell a product provides a glaring example of this kind of error. It is quite conceivable that the vast majority of the people who watch a popular program are in no way influenced by the accompanying commercial. On the other hand, it is possible that a program with a relatively poor rating may have great success in influencing listeners to buy brand A rather than brand B.

The problem has evolved only because it is relatively easy to determine the number of people who watch a given program, but quite difficult—indeed virtually impossible—to determine the percentage of listeners who will ultimately go out and buy the advertised product. In a nutshell, there is a great tendency on the part of statisticians to regard as important those areas which lend themselves to quantification.

Put another way, the unmeasurable or intangible aspects of a situation in many cases may well be more important than those which have been or are capable of being measured. Nevertheless, because the latter are readily available, they become the basis for decision making.

The problem of the unmeasurable often crops up in statistical measures of social conditions. Thus, statisticians for years have been seeking a comprehensive index of the "quality of life." The odds are that they will continue to seek in vain because you do enormous violence to individual values when you try in one comprehensive index to capture a great many different things. In short, any attempt to bring together the very things that make up life—social life, economic life, and political life—is doomed to failure.

Another problem in the social area is the lack of neat statistical relations. The nice thing about a business indicator—say, a price or production index—is that it is part of a statistically related economic system. It is more or less understood that there is a bag of tricks that can be brought to bear to affect these yardsticks. For example, reduce labor costs or demand, and sooner or later prices are bound to fall.

On the other hand, things are not that neat in the social sphere.

Spending more money on education, for example, does not necessarily result in higher educational attainment. Similarly, the link between spending money on police and reducing the uniform crime index (UCI) is definitely lacking. In fact, the result could be quite the opposite, for an obvious reason. If there are more police, they will uncover more crime, and ergo the UCI will be higher.

Moreover, such social indices as the UCI can be easily manipulated for political reasons. Thus a mayor may suggest to the police chief that he would like to show a decline in crime, and as a result a good deal of crime will be "canned"—not shown on the police blotter. This is why the use of crime indices for intercity comparisons or trend analysis is fraught with danger. As a policy tool they are very nearly worthless.

All this is not meant to suggest that social statistical indicators do not have an important public function. All that is implied here is that they do not connect up to a set of decision-making models that automatically can bring into being correctives for undesirable directions of these social indicators.

A related problem of comparability—and one that isn't necessarily limited to the social sphere—involves the self-servicing concept of setting up a norm and then judging data from that. In transportation, for example, we may decide we have to upgrade facilities in order to funnel 4 million people into New York City daily—instead of, say, 3 million. This in turn calls for better roads, subways, railroads, etc.

Perhaps the overall premise is wrong. The more useful answer may well be to stop the New York worker inflow at 3 million—thus forcing private industry to build in the suburbs so that New York people will be funneling out. More to the point, by setting up a norm and then comparing actual performance with that norm, we may well be doing things the hard way.

Perhaps the best explanation of the pitfalls inherent in attempting to measure the unmeasurable was given a few years ago by Arthur M. Ross, Commissioner of Labor Statistics under the Johnson administration. Using scores on an intelligence test as an example, he distinguished between the test (a mechanical instrument) and intelligence (a nebulous concept at best). He pointed out that while scores on a test can be calculated relatively easily, that which is purportedly being measured (intelligence) remains an abstraction. He concluded: "Educators have been learning the painful lesson that great care must be used in drawing inferences from one to the other."

A similar problem exists in measuring some of the more common forms of business and economic data. Thus when one measures inflation, poverty, or the hard-core unemployed, one is measuring man-made, socially defined concepts. It follows then—as in the case of the intelligence test—that people who interpret such measures are often too willing to regard partial or statistical truths as objective realities.

By confusing shadow with substance, they are able to deceive themselves. Such statistics, at the cost of vast oversimplification, enable them to think they have the answer to evasive, many-sided problems. The difficulty, of course, is that the unmeasured or unmeasurable part of the problem is usually far more important than that part which is measured.

Again drawing upon Arthur M. Ross's observations: "These margins for misjudgment are not always stressed to the policy maker. Attracted by the appearance of objectivity and precision, he keeps his eye fixed on charts and tables which may be incomplete, obsolete, or both. Eventually he may come to believe that poverty really is a condition of having less than the current cutoff point of $3,335 in annual income, that full employment really is a situation where the national unemployment rate is 4 percent or less, and that Vietnam is a matter of body counts and kill ratios."

Experience with statistical measures of poverty in the 1960s can perhaps best serve to illustrate the magnitude and complexity of the problems involved in relating an incomplete statistic to action-oriented policy decisions. At that time, as a first step in the drive to eliminate poverty, the statisticians were asked to quantify the definition of what constituted poverty. In doing so, of course, they implicitly set up the dividing line between those below the poverty level and those above.

Over several years, on the basis of this figure (and adjusting it for higher living costs), it appeared that poverty was gradually being eliminated. Yet at the same time the militancy of the poverty-stricken paradoxically seemed to be increasing. What went wrong?

Very simply, the problem was that a simplistic cutoff figure (albeit a useful statistic for some purposes) was inadequate for separating—sociologically speaking—the poor from the nonpoor. The real point was that in the broader social sense, poverty is a sense of failure and despair and a lack of opportunity, good housing, good schools, police protection, etc.

Define poverty in this broader sense, and many of the poor who had jumped over the statistical dividing line were still in the poverty class. In short, dollar income as a gauge of poverty proved to be woefully inadequate.

Another factor that was ignored was the relationship of poverty and the rising affluence at middle-income levels. This pushed the statistical poverty point further and further below the middle-income level. Ergo, there was more discontent among the aspiring poor.

Before leaving the problem of measuring poverty, it might be well to note that even on pure statistical grounds, official unemployment statistics underestimate the seriousness of the problem. They fail to count those who have finally given up after years of frustration and dropped out of the labor market altogether. And they fail to count those who are so poor and so transient that they have never appeared on a census list and hence have never been polled by the government.

Differing Levels of Comparison

As noted above, one prime source of error is the tendency to take a price at one level of trade and apply it to another. Thus when food prices on the farm began to fall in the 1960s, this was widely hailed in the press as a sign that cost-of-living hikes were beginning to decelerate. Subsequent figures, however, showed this not to be the case.

Again the miscalculation could be traced to a comparison of the incomparable: specifically, farm prices and supermarket prices. The correlation between these two broad-gauge indicators was tenuous at best. Indeed, a recent House Agriculture Committee study proved this beyond a shadow of a doubt.

The report showed that the increase in food prices over the past decade has gone mostly into expenses like advertising, packaging, and marketing—activities unconnected with the wholesale price that the farmer gets for his corn or his cows.

The committee report said that two-thirds of the American food bill now goes to the middleman—the companies that process, package, and sell goods—rather than to the farmers. It pointed out that United States consumers' food bills in 1949 brought $20 billion to farmers and $25 billion to middlemen. Today, $30 billion goes to farmers and $60 billion to middlemen. Nor is there necessarily anything nefarious in this trend. The middleman can rightfully insist that the rise in his share of food prices is dictated largely by the housewife's own demand for convenience foods. Frozen french fries and instant mashed potatoes are more expensive to prepare than a sack of Idaho reds.

Errors stemming from making comparisons at different levels, of course, need not be limited to price. It is equally foolhardy, say, to weigh earning performance of manufacturers against that of wholesalers, for each level of trade has its own marketing yardsticks which make sense under the particular set of circumstances involved. Thus, to say that retail profits are unsatisfactory because they are generally under those of manufacturers is patent nonsense. Such statements ignore the reasons for the difference—higher manufacturers' risk, higher dollar investment per dollar of sales, etc.

Still another form of misleading comparison involves the confusion over market and list price. Anyone, for example, who uses list price as the basis for the purchase of a new auto is being taken. Very seldom do marketplace price patterns follow list patterns. The controversy came to a head a few years ago following a Federal Trade Commission report on the diverging relationship between these two car-pricing concepts.

Specifically, the survey found that the manufacturer's suggested price on the window sticker bears little resemblance to what the customer actually pays. A study of 6,585 brand-new 1969-model cars showed that only 1.6 percent of all the sampled American-made cars were sold at the sticker price.

The study further disclosed that more than 70 percent of these cars sold at discounts ranging from 10 to 20 percent off the sticker price. Indeed, it was found that most buyers got at least $200 knocked off the suggested list price; only 11 percent of the customers saved less than $200 (see Figure 13-1). After disclosure of these figures, one con-

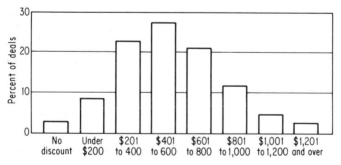

FIGURE 13-1   *How auto sales are discounted* (1969 *FTC study*).

sumer-oriented spokesman suggested that the 1958 sticker law be amended to set standards for establishing the suggested retail price. Such standards, he said, should ensure that the sticker prices approximate the normal selling price for each type of vehicle in a given geographic area.

The dealers and manufacturers didn't like that approach. They contended that to force reductions in the sticker price would remove competition from car selling and that it would limit a dealer's flexibility to compensate for varying factors in each sale—the season, the credit risk of a particular customer, the fact that a particular car is a hot item or a slow seller, or the deals offered by rival dealers to that customer. In short, the auto representatives argued that supply and demand plus the difficulties of valuing the trade-in made the then existing system the only one that was likely to work.

And so the controversy continues to rage. But again the lesson is clear: There always exists the possibility of misinterpretation or outright deceit where two separate market levels of measurement are involved.

Finally, unless great care is used, the "different level of measurement" error is likely to crop up where company performance is compared with general business performance. An example might best illustrate the pitfalls involved.

A great many metalworking firms tend to keep their eyes on durable goods new orders as a barometer of how well they themselves should be doing. They neglect to note, however, that this overall business yardstick is a mixed bag containing a combination of many different order patterns. If a machine tool executive, for example, notes that his 5 percent order rise is under the durable goods 10 percent advance, he may feel something is amiss.

But in actuality his firm may be doing very well—as he would see if he took the trouble to compare the company's performance with that of other machine tool builders. Indeed, if he looked at the machine tool industry ordering pattern, he might find that the industry had suffered a 2 percent decline in bookings over the period being considered. In other words, the worried analyst has been guilty of comparing apples (machine tools) with a mixed bag of fruit (metals, machinery, transportation equipment, autos, etc.). The durable goods order figure, then, is hardly the kind of datum from which to draw meaningful conclusions—it's too much of an aggregate to be used for company benchmark purposes.

Nonrecurring Factors

Many times comparisons are made invalid by special nonrepetitive events. These are often capable of raising or pulling down a statistic for a short period of time without essentially affecting the basic underlying trend. Such hidden influences were discussed in some detail in Chapter 11. At this point the aim is to stress that unless one is aware of these special influences, erroneous or misleading comparisons are very likely to result. Such mistakes are usually due to such factors as the following.

Price hedging.   Let a company announce a price boost scheduled for a month hence, and orders over the next 30 days are almost sure to experience a sharp temporary spurt—as buyers rush in to beat the announced hike. This kind of hedging can be set off even without an official announcement. For example, if a vendor has just been hit with sharply higher wage or material costs, it is a pretty safe bet that he may attempt to jack up prices in the near future.

In any case, to compare sales during such a price-hedging period with sales during a more normal period is an open invitation to error. However, it should be noted that in recent years, the incidence of price hedging has tended to diminish. That's because of today's high inventory carrying charges. Since these costs average anywhere from 2 to 3 percent a month, it hardly pays to lay in an extra month's supply unless the expected price increase is considerably higher.

The same hedging force is sometimes evident in money markets. In 1969 during a severe credit squeeze, banks let it be known that the prime borrowing rate might have to be raised. The predictable result: Borrowers flocked to the banks in droves in order to get in under the wire at the old rate. In six short weeks New York banks alone had to lend nearly $600 million. Most of this represented not the normal borrowing but a hedge. Hence it could scarcely be used as a barometer of businessmen's true money needs.

Strike hedging.   The protection impulse is also usually set off when work stoppages threaten. Thus in the postwar period it has become almost a ritual for steel buyers to stock up every third year, when steelworker contracts come up for renegotiation. This, in turn, has resulted in a temporary bulge and subsequent drop-off in steel inventories—and in a similar pattern in mill sales and profits.

Thus in appraising the steel scene, great care must be taken to separate out that part of a spurt which is due to production needs and that part which can be traced to hedging. It makes year-to-year or even month-to-month comparisons fraught with danger, unless one is aware of the then existing inventory-hedging phase. In 1968, for example, during the height of one of the buildups, manufacturers built up their steel stockpiles to a hefty 15 million tons. A year later these same stockpiles had dwindled to only 10 million tons. This was all due to hedging, inasmuch as consumption (a measure of underlying inventory needs) remained virtually constant over this period.

From a buyer's point of view these sharp buildups and subsequent pare-downs provide necessary insurance. He rightfully figures it's a small cost to pay to avoid costly stockouts in the event that a strike actually materializes.

Nevertheless, these sharp fluctuations (and they are by no means limited to steel) are hardly the most efficient way of doing business, and there have been many suggestions for removing or diminishing strike uncertainties. One, for example, would have new contracts negotiated well in advance of expiration dates. Many feel that only then can expensive and industry-distorting hedge-buying cycles be eliminated. But so far little progress has been made toward this sort of solution—mainly because of the political realities of the power struggle between unions and management.

Legislative pressures. Expectation of government moves which affect business performance can also distort statistical results. A perfect example of this occurred back in early 1969 when the government let it be known that it was considering the elimination of a 7 percent tax credit on new equipment. Since this meant that most firms, in effect, might have to pay 7 percent more for their machinery, the natural inclination was to rush in and place orders before Uncle Sam made his move.

Statistics bear this out. During April of that year (when rumors of imminent elimination of the tax credit were spreading), machine tool orders spurted by a whopping 68 percent in one single month. Again it would be clearly unwarranted to attribute this kind of jump to any underlying need for new machine tools. And again subsequent events proved this out. Indeed, by year-end machine tool orders were running well below even the prespurt levels.

The reverse situation had occurred a few years earlier during a temporary suspension of this tax credit. Then buyers were expecting imminent reimposition. Thus they were holding off in anticipation, and as a result orders dropped off to very low levels—only to spurt again when the long-awaited reimposition finally got the green light.

A rise or fall in income tax rates has a similar distorting effect on profit statistics. For example, when it is known that tax rates are scheduled to fall from year 1 to year 2, many firms will push as many costs into year 1 as possible—thereby lowering profits for year 1 and raising them for year 2. By doing this, of course, they pay a lower overall tax bill over the two-year period. In many cases the savings can be significant.

Note, too, that legislative moves need not always be on the federal level. Just a few years ago several states passed laws making the turning back of car odometers a criminal offense. The net effect was to discourage this procedure—and hence to change the mix on used-car reported mileages. Specifically, with honesty forced on dealers, the number of used cars with low mileages suddenly dropped—and that of cars with high mileages spurted.

The supply "shift," in turn, resulted in a sharp price rise for good used cars and an equal decline for so-called dogs. In short, the time-tested price spread between these two basic types of used cars was abruptly widened by this one legislative action. Indeed, six months after tampering with the odometer became a criminal offense, the spread between good and poor used cars had nearly doubled.

Acquisitions.   Individual company statistics can often be distorted by mergers or the purchase of an additional operating unit or two. The resulting addition to sales or profits makes comparisons with the previous year's performance extremely risky unless the analyst is aware of the changed corporate mix. The best way out of this dilemma, of course, is to compare performance of the same operating divisions over the two periods.

A similar situation crops up when a firm decides to sell off an operating unit. More often than not, this is prompted by the unfavorable profit performance of the division involved, so that while an uncritical look at before-and-after sales and profit totals might suggest a decline, in actual fact that company may have strengthened its underlying financial position.

Changing accounting methods.   A few years ago many manufacturing corporations felt under pressure to report better earnings, owing to the fear of tender offers and takeover bids. Others, engaged in acquisition programs, were eager to show improved profits because they believed that this would make their stock more acceptable when offered in exchange. The predictable result was a wholesale shift to adopt less conservative accounting methods than were then being employed, in the hope that this would result in a faster apparent growth.

One resulting change—a shift from accelerated depreciation to straight-line depreciation for reporting purposes—in some cases made for tremendous differences in apparent profit performance.

Republic Steel, for example, in late 1968 noted that because of this change, net income for 1968 was not strictly comparable to that for 1967. Specifically, the firm said that the new straight-line depreciation had yielded for the first nine months of 1968 a net earnings figure of $65 million, well above the $50 million of the comparable 1967 period. But Republic then quickly added that $9 million of this increase was due solely to the accounting change. In other words, only $6 million out of the $15-million profit gain was due to real factors. The rest was little more than accounting sleight of hand and had to be subtracted if any meaningful comparison with the previous year was to be made.

Sometimes changes in accounting rules can affect one specific industry. Bank income is a case in point. In 1969 banks had to start reporting a net income that included (1) operating earnings, (2) loan losses, and (3) gains or losses from securities transactions—a big change from the previous reporting of just operating earnings. One source estimated that the new rule had the overall effect of reducing earnings about 13 percent.

Generally speaking, firms tend to fight any proposal that will show reduced earnings. In 1968 the American Institute of Certified Public Accountants advanced three proposals to standardize corporate reporting—all of which would have reduced reported profits. Predictably, all three proposals were beaten down.

The way in which firms figure inventory carrying costs is still another accounting-related problem. Each firm seems to have its own formula for calculating these important costs—and hence any comparison between firms is automatically suspect.

The standard textbook way to figure these expenses is to take the cost of borrowed capital plus the cost of maintaining an inventory facility. Unfortunately, not everyone figures it this way. One firm, for example, uses a figure halfway between the cost of borrowed money and the alternative uses of funds. It does not consider facility costs significant because they're a fixed cost; the only time such costs would be considered would be when the firm was expanding. Still another firm emphasizes the rate of return on alternative uses of funds—ignoring the cost of borrowing money since it feels this is implicit in the rate of return on alternative uses.

With this kind of scattered pattern, of course, it is virtually impossible to work up any meaningful industry statistics on inventory carrying costs. It's therefore no coincidence that the government, which seems to have statistics on virtually every aspect of business, has steadfastly steered clear of this controversial area.

Other one-shot factors. Here acts of God play a considerable role. A drought in wheat-producing regions or a freeze in Florida's orange-growing belt may make a serious dent in a given year's production. But again, to assume any underlying shift on the basis of these non-economic factors is to invite erroneous and misleading conclusions.

These one-shot factors, of course, need not be limited to the food area. A war scare may temporarily force up the price of sensitive commodities such as copper or silver. But history has shown that these prices quickly fall back to normal supply and demand levels once the dangerous situation has been resolved. In a related area, the Suez crisis of 1957 (when the Canal was first knocked out) resulted in a hefty increase in United States exports of petroleum. The spurt again was temporary—and exports quickly tailed off the following year. Finally, strikes (aside from the hedging effect noted above) can play hob with reported business statistics. For example, in 1958, during the big steel work stoppage, steel mill shipments dropped by a startling 25 percent. Repercussions were felt in steel-using industries—as well as in the overall United States index of industrial production. But again there were few lasting effects. By late 1959 all industries had recouped their losses and were again back on their established trend lines.

Definitional Differences

In a surprising number of instances comparability is impaired because the two or more figures being studied simply do not represent the same phenomenon. More often than not, the data being compared still go under the same name or title—but that's where the similarity ends. A few examples culled from the field of business and economics will suffice to illustrate the point.

Pay rates. These can be defined as either (1) the actual weekly wage or salary check received by the worker or (2) wages and salaries plus fringes. Since these fringes have been growing at a faster rate than dollars-and-cents pay, their omission generally results in an understatement of labor costs.

According to the U.S. Chamber of Commerce, dollar wages taken alone in 1969 accounted for less than three-quarters of management's true labor costs. The fringes (social security, unemployment insurance, life insurance, profit sharing, etc.) at that time averaged about 98.3 cents per hour, or 27.9 percent of the average hourly pay. More important, this percentage had risen sharply—from 17 percent two decades earlier. Put another way, the percent increase in pure wages over this 20-year period was substantially under the comparable percent increase in the wage-plus-fringe total. The latter, of course, would be the correct figure in computing true labor costs.

The tendency to forget fringes can also lead to erroneous conclusions about purchasing power. In 1968 and 1969 government figures of dollars-and-cents wages—when corrected for a rising price level—suggested that worker purchasing power was falling. And much was made of this conclusion. What was forgotten was that the apparent shrinkage was more than offset by increased employer contributions to health and welfare plans and other such fringe benefits. Indeed, a careful adding of all types of employee compensation (wages plus fringes) would have shown a small increase in real purchasing power over this period.

Inventories. Definitional problems have an equally important potential for error in this key area. Indeed, anyone who has worked in government statistics on business inventories knows that two separate series exist. And surprisingly enough, both are prepared by the Commerce Department. Yet the two do not always move in tandem. One set of figures gives the *level* of inventories at the end of each month. The other series is the inventory component of the GNP which measures the *change* in business inventories and is published quarterly.

Unfortunately, the business analyst simply cannot compare the levels of business stocks for consecutive months, calculate the changes, and arrive at a figure comparable to the change in inventories as recorded in the GNP accounts. There are two reasons for this: (1) The coverage of data is not the same, and (2) the monthly series is at book value, whereas the GNP change is valued at current prices.

On the coverage front, Table 13-1 summarizes the differences between the two inventory series. The differences are with wholesalers, farmers, and the miscellaneous category of "all other." Only merchant wholesalers are included in the end-of-month series, and this series does not include farm inventories or the all-other category.

TABLE 13-1   *Inventory Coverage Differences*

	GNP change in business inventories series	End-of-month inventory level series
Manufacturing	All manufacturing	All manufacturing
Wholesale trade	All wholesalers	Merchant wholesalers
Retail trade	All retailing	All retailing
Farm	Included, but shown separately	Excluded
All other	Included, not shown separately	Excluded

SOURCE: U. S. Commerce Department.

Values of end-of-month inventory and the GNP valuation differ also because of the way businessmen keep their books. GNP, and therefore its components, includes expenditures only for *current* production measured in *current* prices. Businessmen, however, usually calculate inventory holdings on the basis of some mixture of current prices and historical costs.

The following example illustrates the difference: Assume that a firm has 100 physical units of inventory on January 1, 1970. Each unit is valued on its books at $6, making a total book value of inventory of $600. Then shortly after the beginning of the year the price of these units on the open market is raised to $9 each. Furthermore, during the course of the year the firm uses up 80 units and replenishes its stocks with 70 units at the higher $9 price. In terms of current production measured at current prices (GNP basis), a $90 inventory liquidation occurred in 1970 (10-unit reduction valued at $9 per unit).

But the net change in book value on a FIFO basis is positive and is calculated as follows: an additional $630 (70 units added at $9 per unit) less a withdrawal of $480 (80 units withdrawn at $6 per unit). The net effect is a $150 inventory addition.

Thus in terms of book value an inventory accumulation of $150 took place in 1970—rather than the previously calculated $90 liquidation. In a sense, both sets of figures are valid, as long as the analyst is aware of the different concepts used in their calculation.

Balance of payments. In 1967 a major paradox involving international trade statistics emerged. Use of one measure (the "official settlements" basis) showed improvement in the critical United States balance of payments, while use of another (the "liquidity" basis) showed deterioration. Thus, depending entirely on the yardstick used, the payments situation could be regarded as in very good shape or in very bad shape. And there was no official or unofficial rule saying that one definition was better than the other.

The differences between the two concepts, which involve complex financial flows among nations, are beyond the scope of this book. But the fact that two such seemingly similar yardsticks could show such different results points up the necessity of delving into the makeup of any statistical series before any meaningful conclusions can be reached.

While on the subject of international transactions, it may be pointed out that even such seemingly uncomplicated measures as imports and exports can also create comparability difficulties. Again two sets of Commerce Department estimates are involved: (1) census data and (2) Office of Business Economics data. Specifically OBE adjusts the census trade data to conform to balance of payments concepts—mainly by excluding Defense Department foreign sales and purchases, which are reflected in other sectors of the balance of payments. It also includes trade in nonmonetary silver and gold and trade of the Virgin Islands with foreign countries—items which are not in the census foreign trade data.

Thus import-export figures can differ substantially. For example, the 1968 trade balance reported by the census came to $726 million. The OBE figure, adjusted for balance of payments calculations, however, came to less than $100 million.

Statistical Revisions

These are still another headache and often tend to reduce or hinder comparability. Revisions are usually of two kinds. The first involves the release of a preliminary figure based on only partial returns—and the release of a final figure somewhat later when more complete data are available. The preliminary estimate for industrial production, for example, is usually first made available about 15 days after the month being measured. But the more accurate revised figure isn't released for another 30 days.

A second kind of revision occurs where benchmark data have been changed. This second kind is by far the most important and the one that causes most statistical headaches.

First a few words on why these benchmark changes occur. Most statistics—particularly those involving an entire industry or those covering national data—are based in large part on samples. The government, however, will take a census every few years to get a 100 percent head count on a particular series—say, sales or unemployment. In the intervening years, however, samples are taken—and the changes implied by these samples are used to come up with published estimates of the entire population. After a few years errors stemming from these sampling estimates tend to accumulate. These are then corrected, or revised, when the next 100 percent census enumeration occurs.

Statistical analysts view such benchmark revisions with mixed emotions. On the one hand, the better information they provide can permit more meaningful analysis and interpretation—and that's a major plus. On the other hand, there's always the possibility that these revisions may escape the analyst's attention. If they do, he may be comparing current data with erroneous past estimates—and hence may draw incorrect conclusions as to growth and trend.

An equally troublesome aspect of revisions is that the changes often require extensive recalculation on the part of the analyst. Thus, a system of business equations used to calculate, say, sales of autos may have to be completely redone if revised data on disposable personal income become available. With the advent of the computer—and its wider use and manipulation of available data—this recalculation problem often becomes a king-sized headache which affects not only computer calculations but also revisions in computer data banks.

In a few instances, revisions are made not in the old data but rather in the actual collection techniques. A new way of gathering the statistics, for example, will impair the comparability of a series. A recent example in construction is a case in point.

In January 1969 the Commerce Department reported that construction had rebounded sharply after an abrupt December 1968 decline. But in releasing the figures, the agency warned that the movement might not be particularly meaningful because of major changes in the statistical collection of public-sector outlays.

This involved a switch to the collection of public construction data on a monthly basis; previously, reports had been received quarterly from state and local governments, and the government had made its own distribution of a portion to each month on an estimated basis. The change, said the Commerce Department, had caused doubt about recent month-to-month movements—particularly the comparison of the then current estimates with estimates of the previous few months.

While this is not strictly a revision problem, the statistical analyst is often faced with a related dilemma: an apparent inconsistency when comparing two closely related series. Thus, the uninitiated may be thrown by the fact that industrial production often tends to fall when the GNP is still rising. A closer examination, however, would show that the GNP—since it contains (1) services, which tend to resist downturns, and (2) an almost constantly rising price factor—must invariably show a bigger increase (or smaller decrease).

Unfortunately, not all inconsistencies can be explained away this easily. How, for example, do you explain a rise in unemployment when the number of people on the nation's payrolls is also increasing? Some of this, to be sure, may be due to statistical sampling error. But usually more detailed analysis will reveal a good reason for the apparent discrepancy. In the case of the unemployment-employment paradox, it may well be that the labor force jumped sharply one month and that increased employment was able to absorb only a part of this increase.

Structural Shifts

In a few instances, comparability is impaired more by dynamic changes within the business or economic community than by actual statistical shortcomings. In Chapter 12, for example, it was pointed out that a jobless rate which might have been considered low a decade ago is thought to be intolerably high today. In short, straight comparison of the old and the current figures would be incapable of measuring progress or deterioration in the fight against unemployment. A few other such examples of how structural shifts can lead to misleading comparisons follow.

The changed meaning of operating rates.   Some years ago the steel mill ingot utilization rate was used as a barometer of how close to capacity the bellwether steel industry was operating at any given time. A 90 percent rate meant that the industry had the potential of turning out 10 percent more steel ingot and hence 10 percent more steel. But any such interpretation today would be wrong on two counts.

For one, ingot capacity is no longer an accurate yardstick of potential finished steel output. Today ingots usually have to go through three or four further fabricating steps before they become a finished steel product. A bottleneck in any one of these refining operations may well reduce finished steel potential. Ergo, a true measure of steel potential today must include not only ingot-making facilities but also the facilities available in all the other intermediate fabricating steps.

That wasn't, of course, the case 30 years ago, when steelmaking was a relatively simple technique, consisting in the making of ingots and then the utilization of these ingots in finished steel products, with very little additional fabrication. Under this earlier, relatively simple industry setup, the ingot operating rate was indeed an effective measure of finished steel capacity. In short, a good measure of production potential 30 years ago has become a questionable one today because of technological changes.

But operating rates are suspect on another, related count. Some 30 years ago technology moved very slowly, so that a typical mill might be economically viable for 10 or even 20 years. But today—with the rapid shift toward automation—a virtually new plant can be made economically obsolete in a matter of a very few years. Such a mill would still be counted in capacity totals, but because of cost considerations, it would be virtually ruled out as a current operating facility except in cases of dire emergency. Thus to count such a plant as useful capacity could lead to a misleading interpretation of the ability of the industry to turn out steel at existing market price levels.

Steel has been used as the example here, but it should be emphasized that the same basic criticism applies to other industries as well. In short, operating rates are no longer the useful tools they once were—unless, of course, the figures (1) consider all pertinent intermediate stages of production and (2) distinguish between economic and noneconomic capacity.

Selling-method shifts. How a product is procured and who does the buying can influence reported statistics. Plant and equipment is a case in point, particularly for computer purchases. Specifically, (1) the shift toward leasing and (2) the rising volume of industrial bond financing by state and local governments have distorted the true picture.

First consider leasing. When computers are leased by the manufacturer, the cost of production is reported by the manufacturer as a capital expenditure. This cost of production, of course, is considerably under the value that would be counted as capital spending if the computer were sold outright. Because of this factor alone it has been estimated that capital spending on computers has probably been understated by over $1 billion annually in recent years.

Moreover, business capital spending has been understated in recent years because of a rising volume of industrial bond financing by state and local governments. This type of financing was almost negligible a decade ago, but is now at an annual rate of over $1 billion. This represents a substitution of government for private purchasing. Hence, the private capital spending figures reported are lower than they would be if the financing mix were the same as in the 1950s.

Coming back again to the subject of leasing—the trend toward this selling method has also had a major effect on computer manufacturer profits. That's because rentals, unlike outright sales, have the effect of decreasing current income while at the same time increasing income over future periods. In short, changes in the rental-sale mix can result in significant fluctuations in income between reporting periods.

Thus IBM in 1969 issued these words of caution: "Should there be a marked decline in outright sales in a future reporting period, income comparisons would then appear to be abnormally unfavorable, perhaps even to the extent of reflecting a decrease in income for the period affected."

The company spokesman added that shareholders ought to be alerted to the possibility of such a dip in outright sales and profits, despite the fact that orders and production were continuing at a high level.

Measuring a price rise. The growing emphasis on the quality of life (i.e., a cleaner environment) raises some interesting questions about the future adequacy of price comparisons over time. Specifically, today there's an added reason to raise prices: to recoup costs incurred in cleaning up the nation's air and waterways. But does this really mean that inflation has intensified? For is not the consumer now getting more for his dollar in terms of an improved environment? In short, can we compare yesterday's prices with today's prices in view of the "something extra" we are getting today?

Consider, too, the case of many of the public or social services encompassed under environmental improvement, such as a reduced crime rate. There is no easy way to incorporate the value of such improvements in the quality of life into real GNP. So it all tends to show up in terms of a higher price level.

This, in turn, raises questions about government price strategy. If public policies have as their objective a level of GNP price inflation that was appropriate before this shift in priorities developed, there is the danger that efforts to achieve that objective in the face of a steadily growing upward bias in the index could lead to unwarranted restraints upon economic growth.

There are two ways to deal with this. One approach might be for the statistician to just lift his estimate of what is tolerable inflation—say, from 1 or 1½ percent to 2½ percent. But for a variety of reasons, not the least of which would be to put inflation into proper and meaningful perspective, it probably would be preferable to redefine basic price measures through the process of assigning arbitrary quantitative values to qualitative environmental improvement. In a sense this would be a type of quality adjustment—similar to the ones described in Chapter 11.

The usefulness of economic indicators.   Up until a few years ago, the percentage of business yardsticks (new orders, inventories, etc.) turning up or down was regarded as a surefire prediction of future business activity. This indicator technique worked well in the 1950s, but its accuracy gradually began to diminish in the 1960s—despite increasingly sophisticated mathematical processing. The reason for this diminishing reliability was that these indicators cover a gradually shrinking portion of economic activity.

For example, most of them exclude the public area, which now accounts for nearly 25 percent of the GNP. Specifically, the indicator method has little sensitivity for picking up future business shifts induced by the growing role of monetary and fiscal policy.

Moreover, these indicators do not take account of the service sector of the private economy. While it is true that this sector is not yet overwhelmingly important, it is substantial. Consumer spending on services, for example, now tops consumer spending on food and other nondurables. And the bigger this service sector becomes, the less useful are business indicators, which are generally geared to only the manufacturing sector of the private economy.

The changing labor mix.   The heavy emphasis on services is also having major implications for labor. First consider the extent of the shift. Employment by service organizations, ranging from retail stores to care for the aged, now accounts for some 60 percent of total United States employment. In the early post-World War II years, service organizations accounted for only 45 percent of the total.

Moreover, more manufacturing jobs are becoming service-type jobs. In 1950, about 25 percent of employees in manufacturing held service-type jobs, such as selling or doing clerical work. Today, the percentage is approaching the 40 percent level. The same pattern shows up in mining and agriculture.

The rise of service-type jobs suggests, among other things, more stable employment in the long run. The number of jobs probably won't increase quite as rapidly during times of general business recovery, but neither will they disappear as swiftly when a recession sets in.

Services' resistance to change is not hard to understand. Goods, especially durable goods, are periodically produced to the point of oversupply. When this happens, as it did recently in the auto industry, production must be cut back until demand again overtakes supply. In the process, factory workers are laid off.

But things are different in services. Oversupply obviously is not a problem in service industries. Services cannot be stored. They cannot be overproduced. Service organizations don't have to worry about working down excessive inventories. In the end, workers in service jobs are less subject to the ups and downs of the general business cycle than production employees.

But while service growth has added a stabilizing note to the economy, it has also ushered in new inflation problems. Again this becomes clearer when goods are contrasted with services. In the production of goods, for example, the inflationary impact of rising pay rates can frequently be softened through the use of more efficient machinery and production techniques. But it is not so easy to soften the impact of rising pay rates in services. The increasing use of self-service in stores has clearly helped offset rising labor costs in retailing. But how do you automate a barber or a bus driver? Moreover, some services, as they improve over the years, necessitate more rather than less human effort. A case in point is surgery. A difficult heart operation that was not possible 20 years ago requires a large team of doctors and nurses.

In any event, it is more difficult to lift productivity among service-type workers than among workers in, say, an auto factory or a coal mine. Accordingly, as more and more of the nation's jobs become service jobs, it will be increasingly hard to offset rising pay rates through cost-curbing productivity gains.

Technological shifts. Sometimes process shifts have a sharp effect on normal consumption patterns. Steel scrap, one of the principal raw materials used in steel production, is a case in point. Rapid changes in technology in recent years have sharply reduced the use of scrap steel per ton of steel. Between 1960 and 1964 scrap consumption by the three major steelmaking processes and steel output showed similar increases, but between 1965 and 1969 there was only a slight increase in scrap consumption, while steel production rose by 11 percent.

Closer examination reveals that this was due mostly to increased use of the basic oxygen-furnace process (which uses relatively little scrap) and decreased use of the open-hearth process (which uses substantial amounts of scrap). Specifically, the open hearth's share of the market dropped from 87 percent in 1959 to just over 40 percent in 1969.

In short, blind extrapolation of the past scrap–finished steel relationship could have led to serious overestimation of scrap demand.

Note, too, that other technological changes have also contributed to the basic shift in supply and demand patterns for scrap in recent years. One of these is the use of prereduced pellets in electric furnaces. Prereduced pellets that have a 98 percent iron content are fed into the electric furnace to be converted to steel, thus reducing the amount of scrap needed. Although this process is in an early stage of development, it has been reported to be economically feasible and would, if used extensively in the years ahead, reduce scrap consumption still further.

Steel scrap provides only one of many examples of the ways in which new industrial developments have played havoc with time-tested relationships. In any event, it is becoming increasingly dangerous to take any past relationship for granted. Whether it be technological shifts, changing tastes, or just different ways of doing business, the problem is the same: New factors are always at work changing time-honored statistical relationships. Thus to believe that the same forces which have been operating in the past will continue to operate in the future is to leave oneself open to serious miscalculation and error. The business graveyard is full of firms that failed to take these dynamic changes into consideration.

FOURTEEN

Other Pitfalls,
Caveats, and Problems

A surprisingly large number of today's statistical headaches stem from focusing on the wrong question. Too often there's a tendency to ask; What is the answer wanted? rather than, What is the answer wanted for? There's a big difference. Emphasis on the former is an open invitation to distortion—either accidental or premeditated. Complicating the problem is the fact that there are so many ways of going off the track—with the actual number of such questionable approaches limited only by the ingenuity of man.

Some of the key do's and do'n'ts—including those involving percentages and comparisons—have been touched upon in previous chapters. But these are only two of many problem areas. Thus the remainder of this chapter will point up some of these other types of headaches along with some suggestions on how to deal with them.

In a great many cases such statistical problems stem from

a lack of sophistication on the part of either the analyst or the user. In other words, a little more knowledge would automatically prevent many of the problems from cropping up. A case in point: The propensity of some people to equate a mathematical relationship with cause and effect. This can lead to ludicrous conclusions. To take an extreme case, egg prices in New York may vary with the rainfall in New Guinea. But clearly anyone who advocated projecting egg prices on the basis of this relationship would be quickly laughed out of the profession.

The trouble is, of course, that not all examples are this unsubtle. Thus lack of sophistication can result in choosing the wrong approach, using the wrong yardstick, or even being unaware that a problem exists. One firm, for example, used a simplistic hand-calculated correlation approach to project sales—unaware that a more sophisticated, useful, and feasible approach existed—a technique that could be programmed on the company computer at little extra cost and with little extra effort. In another instance, an incredibly naïve and expensive testing procedure was used simply because the firm's research staff had not taken the time and effort to bone up on commonly used sampling approaches.

Attempts to achieve the best of all possible worlds can also sometimes lead to statistical complications. A classic case involving this dilemma cropped up a few years ago when a new federal welfare plan was being formulated. Sponsors of the measure originally wanted both a high poverty floor and strong incentives for the poor to work. But upon closer examination, these two noble goals were found to be somewhat incompatible.

Before detailing why, a few words on how the incentive program would have worked may be in order. Under this scheme, when a poor person took a job, his government poverty-level floor payment would have been reduced by a given percentage of his outside earnings. This percentage reduction—for want of a better definition—can be referred to as the *effective tax rate*. Thus if the floor were $2,500 and the effective rate were 50 percent, a poor worker taking a job paying $1,000 would have had his previous relief payment of $2,500 reduced by 50 percent of $1,000—or by $500. In other words, his income after taking the job would have been $2,000 from government relief checks and $1,000 from outside sources—or $3,000 in all. Thus by working he would have been able to end up with $500 more than if he had not worked.

Unforeseen problems, however, arose when attempts were made to raise the floor and keep the effective tax rate low (say, at 50 percent).

Specifically, if the floor had been raised to $3,500, the government would have wound up paying some welfare to any family earning less than $7,000 (for only at $7,000 would the effective 50 percent rate have completely caught up with the new welfare payment floor of $3,500).

But this was clearly untenable and unfair because at the time, the median family in the United States made only $8,000. Clearly, the government couldn't give a subsidy to one family when another family making only slightly more net income received none at all. It took a lot of hard explaining to convince proponents of the liberalized plan of this self-defeating result. A little more statistical sophistication at the outset could have prevented a lot of hard feelings and delay in getting the program off the ground.

But lack of sophistication can't always be blamed for statistical mistakes. In a great many instances outright attempts to deceive are behind the errors and misinterpretations. Examples of this—particularly in advertising—are numerous. But deceit can also appear in more subtle forms, with the reader hard put to separate fact from fiction. The section below on spuriousness points up many such cases.

At still other times, problems arise solely because of the enormity of the task being tackled. Thus despite statistical sophistication and lack of bias, sheer complexity often makes it extremely difficult to arrive at a valid and useful result. The basic problem in this instance, of course, is that the outside world is never as simple as we would like it to be.

A few years ago, for example, a well-known consulting firm tried to forecast price trends by relating the price level of a specific commodity to one or two determining variables. The approach failed miserably—primarily because the firm wasn't prepared to grapple with the complexity of pricing.

Indeed, it was only when the firm realized this—and began to factor in all the pertinent forces influencing a given commodity's price level—that its track record began to improve.

As a starting point, the consulting firm had to realize that while some prices are market-oriented (responding sooner or later to forces of supply and demand), others are cost-determined (raised or lowered by management in accordance with unit cost pressures). In addition to these basic factors, the consulting firm found that price levels are determined by (1) prices charged by competitive firms, (2) prices of substitute products, (3) import prices, and (4) unusual or temporary events affecting the flow of supplies.

The above example illustrates a basic principle of the statistical approach: There is no such thing as instant or shortcut analysis. It just doesn't work—and usually results in more harm than good. A good quantitative analysis takes time. The problem is essentially a threefold one: (1) determining what is needed, (2) estimating probable costs, and (3) determining whether or not the effort is worthwhile in terms of eventual payoff. If it isn't, then it is better to scrap the entire project. On the other hand, if the study seems worthwhile, by all means go ahead—but be prepared to pay the cost. In short, recognize the basic truth that "you get nothing for nothing."

With all the above in mind, here's a rundown on some of the more common types of pitfalls, caveats, and problems not covered in previous chapters.

Spuriousness

In the statistical sense, the term "spurious" is usually applied to figures which are purportedly very meaningful or significant, but which in truth are not. Such spurious statistical presentation can take many different forms, such as those discussed below.

1. Fictitious accuracy. The old "dotting the i" approach is probably one of the most common of these ploys. A figure followed by two or three decimals seems a lot more accurate than the simple rounded-off number. But this need not always be the case. As a rule of thumb, mathematical calculations based on rounded-off original data should be presented in rounded-off form. The reason is simple enough: The end product of data processing can be no more accurate than the original input.

But the human mind tends to forget this rather obvious fact. Tell somebody that 2.145 people out of 10 smoke brand X, and it will seem to him that a lot more research went into that calculation than into one stating that about 1 person out of every 5 smokes the given brand. Yet the research performed may not have involved anything more exacting than the research backing up the "1-out-of-5" statement.

The basic mathematical rule is to present no more than the number of significant digits. More specifically, one must limit the number of digits to no more than those contained in any one of the figures involved in the calculation.

EXAMPLE: If the total income of a firm was $101 million and the number of customers was 102.3 thousand, then to calculate the average sale per customer, one would divide the number of customers into income. Thus $101 million divided by 102.3 thousand yields a figure of $987.2922. The reputable analyst, however, noting only three significant digits, would say that the average customer purchased about $987 worth of goods. Any more detailed figure would be statistically meaningless.

2. Unnecessary accuracy.   Sometimes the degree of precision, even if statistically valid, is not worth the effort. In the above illustration, for example, it is conceivable that both the number of customers and the total income could have been calculated to enough places to yield a statistically valid answer to the nearest cent. But to what avail? Possibly we might find out that the average customer purchased $987.49 worth of goods. For policy decisions this is not a bit more helpful than the approximate $987 figure determined by using the rounded-off figures. So why go to the expense and effort to achieve something which has no apparent payoff?

To be sure, the money involved in achieving the greater-than-needed accuracy in the above example is small. But in other cases it can run into considerable sums. Thus a few years ago a company wanted to forecast the price of several types of steel it purchased on a regular basis: cold-rolled carbon sheet, hot-rolled sheet, and cold-rolled bar. The company had two alternatives: (*a*) It could work up an approximate equation to forecast the average steel price trend, and then from the normal price differential among products forecast prices for each of the three individual product lines, or (*b*) it could work up individual forecasting equations for each product separately—a procedure that in this case would entail about three times the amount of work.

Clearly, the second approach, since its predicting equations would be custom-tailored, would yield somewhat more accurate results. But it could be achieved only at a much higher cost.

When presented with all the facts, the company chose the first, more approximate approach, recognizing that it was more than ample for the use to which the data would be put—and further recognizing that the money thus saved could be funneled into more productive statistical analysis in other areas. In the above case the firm used the money saved to make a similar forecast for another important purchased commodity—copper.

3. Computer spuriousness.   The introduction of the electronic brain, with its ability to perform thousands of calculations per second, has led to some unintentional misuse. An analyst who might want to predict sales of, say, a chemical with the help of a computer can now experiment with literally hundreds of different determining variables until he hits upon the right combination. Or he might use so many of these variables in his predicting equation that the statistical validity of his results could be seriously questioned.

The second danger is one that can be avoided with just a little statistical savvy, for sampling theory clearly spells out the inverse relationship between reliability and the number of variables: The more variables used to explain a given relationship, the greater the chance of sampling error—and hence the less the reliability. Thus before any forecast goes out, it should be standard operating procedure to check for reliability. More will be said on this subject in the next chapter, where the overall question of accuracy is examined in greater detail.

The other danger referred to above—juggling variables until the right combination is hit—also deserves some brief comment. If one tries long enough—and has enough ingenuity and patience—one is likely to come up with a good relationship. The only question is, Just how meaningful is it? The fact that chemical A is correlated with, say, demand for autos may be little more than happenstance—and not any proof of cause and effect. While the whole cause-and-effect problem will be dealt with separately below, it is important to recognize the increasing likelihood that one may try to justify a cause-and-effect relationship when one—after much searching and expense—comes up with a good statistical relationship.

In all fairness it should be pointed out that this "forcing" isn't always premeditated. The propensity to delude oneself is just as strong in the statistician's psyche as it is in anyone else's. The difference here, however, is that the analyst is a professionally trained person—and hence is expected to be more than normally aware of this kind of pitfall. If he isn't, he simply is not doing his job.

4. Advertising distortions.   One need only turn on his TV set to see this in its most blatant form. In many cases little or no sophistication is needed to see through the transparent distortion of facts and figures intended to hammer home a particular advertiser's message. A few of the more common forms of distortion in which statistics are used as a basis for confounding the public are discussed below.

Rigging the results.   A typical example here would be the toothpaste company that wants to show that most young people prefer brand X. The company might distribute free samples one morning and then later the same day hire an independent research outfit to check brands used in the particular geographic location. The almost predictable result is that more consumers will mention brand X because it is still fresh in their minds. In statistical parlance, the earlier distribution of free-sample toothpaste distorted the sample.

Loose phraseology.   This is another useful advertising gimmick. "Zingo detergent is three times better," reads an advertisement. But what does it mean? Three times better than what? The truth of the matter is that the statement has no real meaning—and is made only to leave the reader with a positive interpretation. The same can be said of a statement that one brand of cereal is 30 percent better-tasting than competing brands. Again there is no way of knowing what yardstick (if any) is being used for the comparison.

Similarly, a washing machine may be advertised as the one with the lowest cost. But what does this mean? Is it the same thing as saying that it is the cheapest? Not always—for while the initial outlay may be low, performance may be poor or repair costs may be high. The point here is that cost and value are being confused. Obviously, a cheap washing machine is virtually worthless if it is always breaking down.

As another example of loose phraseology, a tobacco firm may say that its cigarette is lowest in tar or nicotine. This may be true, but the difference between the amounts of these poisons in all cigarettes may be so small as to have little statistical significance. Or if there should be a significant statistical difference, it may be irrelevant—particularly if a small amount of the poison is as injurious as a larger amount.

Unwarranted additions.   Consider the example of a survey which asks a detergent user whether product A yields (1) a cleaner wash and (2) a fresher-looking wash. If 3 out of 10 say "cleaner" and 4 out of 10 say "fresher," it is incorrect to state that 7 out of 10 find that the detergent gets clothes cleaner and fresher.

The reason, of course, is that if a user says the detergent gets clothes both cleaner and fresher, he is being included twice in the tally—once under "cleaner" and once under "fresher." Addition is possible only when the categories are mutually exclusive.

The irrelevancy ploy.   Again drawing upon a cigarette example, consider the popular advertisement of a few years back which stated that more doctors preferred brand X than any other brand. Further, make the questionable assumption that the sample was not rigged. Despite all this, the commercial's conclusion that the average consumer should smoke brand X is still spurious, for it rests on the shaky ground that doctors know more about the comparative health hazards of individual brands than the ordinary consumer does.

This is, of course, patent nonsense. For one thing, it would presuppose that the doctors had done exhaustive research tests on all brands—something that each and every doctor queried couldn't possibly have done. But the conclusion is vulnerable on still other grounds. Since it is generally agreed that cigarettes are injurious to health, why would any doctor—who is presumed to be all-knowing about all types of health problems—be smoking at all?

Built-in bias.   Often advertising will base claims on opinion surveys, which are vulnerable to bias on many counts. For example, ask a group of women their ages, and chances are that the answers will be somewhat under the true levels. The fault here, of course, is in the question itself, which fails to take into account the vainness of some women.

Quantification for the sake of quantification. The numbers game in rating TV programs is another deception practiced by Madison Avenue professionals. Everybody knows it is meaningless, but because of the need for some quantitative data—whether they are right or wrong—ratings have become part of the TV game.

Chapter 5 examined some of the other ways in which bias can creep into opinion surveys. The whole subject is brought up here only to emphasize that this is one of the major areas in which outright distortions and self-deception are practiced with considerable success by men who should know better.

Cause and Effect

This is essentially another—and perhaps the most important—area where spuriousness can creep into statistics. The problem—briefly alluded to above in the discussion of computer spuriousness—is basically this: In doing any kind of statistical work relating to two or more variables, it is crucial to ascertain whether the relationship between the variables or series being analyzed is a rational one—one backed up by a body of theory.

The running controversy over the relationship between cigarette smoking and cancer points up the type of problems involved. There is an undeniable mathematical relationship between cigarette smoking and cancer. But the cigarette industry claims that this does not prove cause and effect. They argue that the factors that lead to cigarette smoking may be the same ones that cause cancer. In short, the smoking itself is of little consequence. If these claims are true, of course, cigarette smoking could be completely abolished and there would be no appreciable decline in the number of lung cancer cases reported to the medical authorities.

In a sense, the cigarette industry is claiming that the covariation of the two variables (cancer and cigarette smoking) is due to a common cause or causes affecting both variables in the same direction. They might say, for example, that tensions cause smoking and that tensions also cause cancer—and that's why these two variables move in tandem.

As another example, this one from the world of economics, many people have noted the inverse relationship between inflation and unemployment—sometimes referred to as the *Phillips curve* or the *unemployment-inflation trade-off*. When unemployment is low, prices tend to rise at a fast clip. Conversely, when unemployment is high, a greater degree of price stability seems to be achieved. This had led some to believe that you can curb inflation by raising unemployment.

A moment's thought will show that this is again pure nonsense. What is happening here is that the influence of a third variable, demand, is being ignored. When demand is high, prices rise and unemployment dips—and vice versa. In any case, unemployment does not cause inflation.

The lesson to be drawn is this: The mere existence of a correlation does not prove cause and effect per se. This doesn't mean that one should not perform statistical correlations to test a causal hypothesis. If one hypothesizes that A causes B, one may test this by correlating values of A and B. If a correlation does indeed exist, it merely supports, but does not prove, the hypothesis.

This procedure is tailor-made for many types of business problems. Take sales and income. Theory suggests that sales vary directly with income—rising when income rises and falling when income falls. A correlation is then computed between these two variables which then supports—but does not prove—this thesis.

All this is not to say that statistical evidence can't be the starting point. In many cases an observed relationship between two variables leads to the establishment of a new theory. Take again the cigarette-smoking–cancer relationship. The cigarette industry was clearly correct in stating that the mathematical relationship does not prove cause and effect. But the relationship did suggest that a connection might exist. This, in turn, led to laboratory experiments proving that certain agents in tobacco could physically cause cancer.

The point to remember is that the real proof was established in the laboratory through clinical tests. The statistical relationship only pointed the way and subsequently backed up the laboratory findings.

Sometimes when cause and effect does indeed exist (theory confirmed by statistical relationship), it is often advisable to amend the basic relationship as new causes come into play. This time consider the relationship between capital spending and its two key determinants—capacity utilization and profits. As pointed out in a previous chapter, additional forces (high labor costs, competition, and obsolescence) have appeared over the past few years. And these, too, have to be factored into the predicting equation.

But over and above such steps, even more precision can be obtained by buttressing the predicting equation with an ancillary relationship. Specifically, by analyzing the relationship between capital appropriations (when the money is first appropriated) and capital spending (when the money is actually spent), we can get a better fix on the exact timing

TABLE 14-1

Money appropriated today will be spent according to this pattern:

Next quarter	5%
Second quarter	20%
Third quarter	20%
Fourth quarter	30%
Fifth quarter	20%
Sixth quarter	5%

of any shift. Table 14-1 spells out this relationship between appropriations and outlays. Note that there is little spending activity the first three months following the initial appropriations—as plans get finalized, contracts are let, etc. Then the percentage of appropriations spent gradually increases, only to taper off, with the final installment completed in 18 months.

One consulting firm uses these lead-lag relationships to zero in on plant and equipment spending—after having determined the more general investment trend by correlating capital spending with its determining variables. In short, they wring every last bit of useful evidence out of statistical relationships suggested by business and economic theory.

Erroneous Comparisons

A few comments on unwarranted comparisons are also in order. Aside from the broad family of comparison-type problems treated at length in Chapter 13, there are some specific distortions that consistently crop up in business areas. They are usually used with an intent to deceive, for the shortcomings in all the illustrations noted below are or should be well known to statistical professionals.

1. *The labor rate–unit labor cost deception.* The usual approach here is to cite rising labor rates as a reason for boosting prices—ignoring the fact that productivity (output per man-hour) may well have offset such hourly pay boosts. The point is that prices are related to unit labor costs (pay rates divided by productivity) rather than to labor rates. It is the former that constitute real costs to any firm.

A simple example can help put this into sharper focus. Assume that at the XYZ Corp., which produces widgets, the only cost is labor. The company's five employees produce 10 widgets per hour at a labor cost of $1 per hour per employee. Further assume that the XYZ Corp. sells the product at $1 per widget. All this is summarized in the first part of Table 14-2, which shows a profit of 50 cents per unit, or $5 on the 10 units produced in a single hour.

Next assume that new equipment permits a doubling of productivity (20 units per hour instead of the previous 10) and that labor, in turn, asks for a doubling of wages ($2 per hour instead of the previous $1 per hour). Management, in an effort to distort, may well point up the doubling of wages, but ignore the doubling of productivity. But look at what actually happens (see the second part of Table 14-2).

The labor cost to turn out one widget remains the same—the doubling of the pay rate offset by the fact that each employee now turns out twice as many widgets. Note, too, that because twice as many widgets are being turned out, profits have actually doubled (assuming, of course, that all the additional widgets can be sold). In short, the 100 percent pay rise when coupled with a 100 percent productivity raise has actually aided management as well as labor.

TABLE 14-2 *The Cost Effect of Productivity and Wage Hikes*

	Before	After
1. Labor costs per hour	$1	$2
2. Widgets produced per hour	10	20
3. Number of workers	5	5
4. Total labor costs (1 × 3)	$5	$10
5. Unit labor cost (4/2)	50¢	50¢
6. Selling price	$1	$1
7. Unit profit (6 − 5)	50¢	50¢
8. Total profit (7 × 2)	$5	$10

The above illustration also destroys another myth—the one that states that if labor is given a wage increase equal to its productivity increase, it leaves nothing for management. As a general rule, the granting of wage increases equal to productivity increases always leaves ample opportunity for an increase in corporate profits. Indeed, this is the thinking behind government's wage-price guidepost theory introduced in the early 1960s under President Kennedy's administration. It stated that wage increases should be in line with productivity gains—and that such a policy would (1) eliminate the need for price increases and at the same time (2) provide higher profits to corporate management.

Putting the emphasis on unit labor costs rather than labor rates also demolishes another popular myth—namely, that high United States pay rates are pricing American goods out of world markets. True, United States pay rates are higher than those anywhere else in the world. But it is equally true that United States productivity is higher, with the result that average unit labor costs in the United States are no higher than those in any other industrial country.

Those who doubt the validity of this statement have only to look at trade figures for further proof. United States exports consistently outrun United States imports. If our wage costs were not competitive, it is inconceivable that we would be selling more abroad than we purchase from abroad. This offers proof positive that America is still competitive in world markets.

That's not to say, of course, that United States and world unit labor costs are equal for each and every product. In some areas (e.g., machinery) our unit labor costs are lower. But in others (e.g., electronic components) other countries have the advantage. But that's what makes a ball game. And in the end—with everybody doing what he does best—the standard of living tends to rise all around the world. Economists, incidentally, refer to this as the "law of comparative advantage" and use it as the basic justification for free trade.

2. International comparisons. While unit labor costs aid in comparing the production costs of various countries, there are other areas where it is extremely difficult to weigh the performance of one nation against that of another. Take international exchange rates, which supposedly permit comparison of the GNP and wealth of one country with the GNP and wealth of another.

The exchange rates do not precisely measure comparable purchasing power among countries. For one thing, these rates are oriented more toward the types of goods entering international trade—and for any given country these may not be the same goods consumed within that country. Second, political factors often dictate an exchange rate above or under the true purchasing-power level. Thus France a few years back was reluctant to devalue her currency because this would be an admission of failure to maintain a favorable balance of payments position.

But even assuming no mix or political distortions, exchange rates are still incapable of measuring comparable purchasing-power levels. Thus the income of poor countries is to some extent underestimated on two counts. First, the staples that form the standard of living of the masses are generally very cheap, compared with the items bought by higher-income families. Anyone who has visited an underdeveloped country can not help but be impressed by the fact that while it is relatively expensive to live "American style," the vast majority of people, by shopping in native stores, can get by on very little money.

Second, the purchasing power of underdeveloped countries is underestimated because so much home-produced food and clothing fails to go through the marketplace. And if it doesn't go through the marketplace, it is not picked up in GNP totals—since GNP is defined as the volume of market transactions. For some of the more backward countries this factor may result in underestimation of purchasing power by as much as 25 or 50 percent.

Omitting the Yardstick

Here the problem lies in drawing conclusions from some statistical finding without taking the trouble to relate the figures to a meaningful yardstick. A few simple examples will serve to illustrate this family of problems.

First consider a survey on automobile accidents where one breakdown shows that more accidents occur during the day than during the night. A careless analyst might conclude from this that it is more dangerous to drive during the day. He would be overlooking, of course, the fact that more people drive during the day. The meaningful comparison would be between the percentage of accidents that occur during the day and the percentage that occur at night—and this in all probability would lead to the opposite conclusion regarding the relative safety of driving during the two periods.

This type of distortion often tends to occur when an argument is presented to put an individual, a firm, or an industry in a favorable light. Thus a statement that one particular industry pays more income taxes than another should always be open to question. In this particular case, the question is, What were the specific incomes of the industries being compared? It would be quite odd, indeed, if the automobile industry did not pay a much larger tax than, say, the leather or cement industry.

A similar oversight occurred a few years ago when the petroleum industry proudly declared that its fringe benefits to workers were running more than $1\frac{1}{2}$ times higher than the average for all manufacturing. What this industry conveniently forgot to point out was that the salaries of petroleum workers were also about $1\frac{1}{2}$ times those of workers in other industries, so that fringes as a percentage of total pay were the same in petroleum as in other industries. Again the lesson is the same: Numbers when not related to appropriate yardsticks are virtually meaningless—and may well be used to distort the truth.

The old national debt myth would fall into this category. Year after year prophets of doom point to the growing national debt—noting that we are surely heading for bankruptcy. What these people neglect to point out is the equally fast or even faster rate of economic growth. In other words, the federal debt as a percentage of growth (a measure of ability to carry debt) has actually been declining. Indeed, the federal debt–GNP ratio is now below levels prevailing at the end of World War II. Those who ignore the pertinency of this ratio approach are in essence ignoring the fact that a $100 debt to a ghetto-dwelling inhabitant is a lot more overwhelming than a $100,000 debt incurred by a millionaire.

Sometimes the establishment of a yardstick is a problem—particularly when more than one choice is possible. The temptation to distort is always great when the use of one of the alternatives tends to put a company or industry in a more favorable light than the use of another one. The alternative profit-margin choices alluded to in Chapter 12 are a case in point. Since earnings per dollar of sales are usually well under earnings per dollar of stockholder equity, the natural inclination of business firms is to stress the former when pleading poverty—and the latter when they want to impress stockholders with the quality of their management.

But aside from deciding which yardstick to use, it is also important to keep in mind that a yardstick will tend to vary industry by industry, firm by firm. Again taking profit margins as an example, a firm or industry with a large turnover will generally tend to have a smaller profit-per-dollar-of-sales ratio than one with a small turnover. But this seldom has any deleterious effect on general profit performance. Note, for example, that the food industry, though notorious for its low profit per dollar of sales, is still one of the most profitable as far as rate of return per dollar of stockholder equity is concerned—primarily because of its high turnover rate.

The choice of yardstick also crops up in productivity measurement. The auto unions, in negotiating with General Motors or Ford, will always cite the productivity gains of the auto industry, which just happen to be the highest or close to the highest in the nation. Management, on the other hand, will cite the all-industry productivity figure (which includes low service productivity). And if labor shoots that one down, management then wheels in a figure on all-manufacturing productivity—a higher measure than the all-industry figure, but still considerably below that for autos taken separately.

Which one is correct? None or all—depending upon your point of view. In short, all statistics can do in this case is to point up what each yardstick measures and then let the parties concerned determine which is the most appropriate for the problem under discussion.

Errors of Omission

In many instances, forgetting to present some pertinent statistical information can be about as misleading as any outright juggling of the figures. In both cases there is an attempt to give a desired impression or achieve a desired result—even if it is an incorrect or dishonest one. Failing to relate a finding to its appropriate yardstick (touched upon above) is only one of many ways in which the analyst can conveniently forget to include some pertinent bits of information. A brief rundown on some of these other types of omission follows.

1. *Accounting omissions.* Up until recently many firms did not include breakdowns on profit and loss by product line. The lack of this and other detailed data hurt not only investors but also customers of the reticent corporations. Such information, for example, enables buyers to know where vendors are making their big profits and where vendors might be able to afford to provide a price break or more liberal selling terms.

Or, looking at it from the reverse viewpoint, if buyers know where a vendor isn't making enough money, they know where he may be willing to work a little harder to make a sale. Then, too, better information about product lines could give a buyer advance warning on any product line that might be in danger of being dropped.

Incidentally, new pressures by Uncle Sam are forcing corporations to make many more useful breakdowns in the cost and profit areas. The entire subject of financial reporting—and how it is slowly changing for the better—is discussed in greater detail in Chapter 10. Interpreted correctly, these breakdowns should lead to more accurate forecasting by recipients of the reports, more realistic market readings, and, as pointed out above, better buys for users of the reporting company's products.

2. *Qualification omissions.* Many reports are admittedly based on less-than-perfect data. Sometimes this is due to difficulties in obtaining accurate information—and at other times to the need to keep costs down. In any case, unless these shortcomings are clearly stated, the recipient of the statistics is likely to place too much credence in the actual numbers—and to move one way or the other with certainty at a time when caution may be called for.

A few years ago, for example, a domestic producer was presented with data on average import prices, which were to help him make a decision on whether to drop a particular line. The data were accompanied by impressive charts—suggesting that the figures were pretty near impeccable. But what the analyst failed to tell top management was that no price index of imports was available and that he had substituted the next-best thing: an index of average unit prices (value divided by the number of units).

Unfortunately, this substitute for a price index is unduly influenced by changes in product mix. Thus a trend toward the importation of cheaper models would lower the average unit price, while a trend toward more expensive models would do the opposite—raise the average unit price.

The data presented to management reflected the former. But management, unaware of this shift in mix, attributed the growing gap between import and domestic prices to a lack of competitiveness and was on the verge of dropping out of the market in question. Luckily, at the last moment the shortcoming of the data was brought out into the open. Management then rightfully concluded that the growing gap was due to the increased importation of cheaper models—and hence decided to continue producing, with the accent on the more expensive models. Subsequent events proved the move out, with the product lines emphasized becoming major money-makers only two years later.

Another qualification often omitted is the range of error. No matter how good the data are, any forecast must vary if only because of chance factors. Any report that fails to bring this fact of statistical life to management's attention does a serious disservice. The actual ranges of error and how they are calculated will be discussed in greater detail in the following chapter.

3. *The selective ploy.* Here the analyst will generally have the statistical evidence that both supports and refutes his major thesis. But in submitting his report to management, he stresses the positive aspects, while playing down or even eliminating the negative ones. This may be a good or even great approach for a salesman, but it is essentially self-defeating in terms of statistical analysis. Company officials are paying good money to get the facts, and it is something less than honest to shortchange them.

A researcher correlating company sales with income and prices of competitors, for example, may find that price does not seem to help explain variation in sales. His first inclination is to ignore this and stress the income/sales relationship. It may well be, however, that the competitive price data he is using are inadequate, or perhaps there is a good economic reason why the prices of competing products do not help explain changes in sales. In the latter case it is incumbent upon the researcher to point up this lack of relationship in his report. Indeed, with this knowledge, management may be more willing to chance a price increase than it would otherwise be. Or, conversely, it may be less tempted to cut prices in order to boost sales.

Averages are also often used to hide useful data. For example, a firm may make much of the fact that its purchase costs remained unchanged during a year of inflation. This is to imply heads-up purchasing. That may, indeed, be true. But then again it may not be. For example, it is possible that the one big item the company buys may have been in excess supply—forcing prices down. This decline in one key raw material may have been enough to offset hefty increases in other smaller-volume purchases.

A realistic report would point all this out and would supplement data on purchasing costs with comments on the trends in the various materials that made up the average.

The news media are also often guilty of omitting pertinent data—usually to heighten the newsworthiness of statistical developments. Let inventories rise a billion or so, and the headlines proclaim the likelihood of an inventory glut. What they may forget to point out, however, is that sales also moved up by about the same percentage, so that the inventory/sales ratio—a measure of inventories relative to need—has not changed at all.

Sometimes even the government is guilty of this selective approach. In early 1970, when the administration was committed to an all-out fight on inflation, government spokesmen stressed the point that the consumer-price-index increase had decelerated from a 0.5 percent rate of advance the month before to only a 0.4 percent advance during the month in question. But they neglected to point out that 0.4 percent was still way too high for comfort and that the small downward change noted may well have been due to sampling errors rather than to any significant lessening of inflationary pressures.

An even more blatant effort at distortion—also involving the rate of inflation—occurred that very same year. When the second-quarter 1970 GNP figures were first released, much was made in Washington of the fact that the annual rate of price rise had decelerated from 6.4 down to 4.2 percent. But hidden deep down in the report was a statement that all the improvement had been due to statistical quirks and a reweighting of the components. The report then concluded (in fine print) that there was absolutely no evidence of a price slowdown. But the White House, anxious to show how its anti-inflation policy was working, conveniently forgot to mention this important qualification.

Underlying Assumptions

Sometimes, with all the stress put on statistical accuracy, a key element of any quantitative presentation is forgotten: the basic ground rules on which the study was based. Many a seemingly perfect forecast has come crashing down because of the shifting sands on which its foundation was laid. Assume prosperity for the next year, and no matter how bullish your firm's outlook is, sales will be lower than anticipated if the economy should go into a tailspin. Actually, two separate problems, discussed below, are involved.

1. Tenable and untenable assumptions. In not a few cases the problem can be traced back to an unrealistic assumption. Since everything may depend on this assumption, it is clear that the final results or recommendations may also be suspect. Again a few examples can best illustrate the problem.

First consider a company that was about to launch a new set of dishware based on plastic X. The company, analyzing the market, made two questionable assumptions, namely, that through intensive advertising it would (1) be able to switch a greater share of the dishware market over to plastics and (2) be able to capture its fair share of that expanded plastic dishware market.

It turned out that neither of these two assumptions was warranted. Many other plastics had previously hit the field and had just about saturated that part of the total dishware market which would be willing to substitute plastic for the traditional, more prestige-oriented china types. Second, those other plastics already in the market were deeply entrenched and had a loyal following. In short, it was a lot harder than anticipated to convince these people to switch from existing plastics to plastic X.

The final result was predictable. Despite accurate consumer purchasing-power estimates and near-perfect cost and production projections, calculated sales fell 55 percent under projected levels. The problem, of course, was that nobody took the time to question the two underlying assumptions—neither of which proved tenable in the crucible of the real marketplace.

Even such self-styled experts as Uncle Sam's prestigious Council of Economic Advisors have been guilty of the unrealistic or incorrect assumption syndrome. The 1968 income tax rise and its projected effect on consumer sales is a case in point. The assessing of a 10 percent surtax that year was expected to make for a rather sharp deceleration in consumer spending gains. But as later events proved, the opposite actually happened: spending accelerated.

Much of the explanation may lie in an array of psychological reactions that couldn't be cranked into the computers. "I think clearing the air by passage of the tax bill had a lot to do with it," suggested one insider. As he saw it, people were cautious about their spending when there were (1) many headlines about the possibility of a heavier tax burden ahead and (2) dire warnings of world monetary crisis if the burden wasn't imposed. Once people could see exactly how much (or how little) their paychecks were affected by the higher withholding, they relaxed and resumed their customary confident shopping. Carried to its logical extreme, this argument has it that the main dampening effect that could be expected from the surtax was the one that took place before it became a reality.

Increasingly, too, strategists in retrospect have reassessed the fiscal (though not necessarily the social) wisdom of sparing millions of low-income families from the surtax. These are the people most likely to spend all they take in. Tax these people more, and they are usually unable to offset the extra tax bite by saving less or borrowing more. Ergo, their purchases go down.

Not so with the middle- and upper-income families, who actually had to pay the surtax. For the most part they were eminently able to offset the extra tax bite by saving less and borrowing more. Ergo, there was little downward pressure on sales totals.

2. Changing assumptions.   In other cases assumptions may be perfectly valid at the time they are made. The only trouble is that things are prone to change as the forecast period approaches. Thus in the late 1960s predictions based on physical need and capability called for the construction of more than two million housing units per year. This assumed that enough money to finance this number of starts would be available.

Unfortunately, it was just about this time that the government stepped up its fight against inflation by making money both tighter and more expensive. This hurt housing, and as a result, starts during these years fell below 1.5 million units and in one year, during the height of the credit squeeze, weren't much above 1 million units.

This, in turn, played hob with long-range forecasts of building material producers. Demand for cement, brick, glass, lumber, and a host of other items fell sharply under forecasted levels. Some of the firms involved were alert to the money shift and adjusted their forecasts accordingly. But others, forgetting to make periodic checkups of underlying assumptions, found themselves overproducing, with the resultant heavier-than-normal inventory buildup forcing a sharp drop in profits and margins.

At other times it is simply an act of God or nature that can turn a particular assumption upside down. Air-conditioner manufacturers, for example, in assessing the near-term outlook always have to make an educated guess about the weather pattern during the summer—their key selling season. Since weather forecasting is more an art than a science, most of these companies factor in the assumption of normal weather. But since normal is derived by averaging the temperature of many different summers, the chances that there will be just normal weather during any particular summer are pretty slim. So most air-conditioner makers, by necessity, have to keep a running check on this assumption—adjusting production schedules several times a year to take into account any sharp variation from the so-called normal weather pattern.

A good example of what this unpredictability of the weather can lead to occurred during the summer of 1966. A simmering heat wave in late June and early July caught the room-air-conditioner industry unprepared—depleting stocks of both dealers and distributors alike. A spokesman for the industry estimated at the time that some 5 to 8 percent of that year's sales potential was dissipated because of the inability to meet demand. The problem here, of course, was aggravated by the fact that demand is seasonal and if it cannot be met at the proper time, it is lost forever—or at least for the year in question.

Another assumption—this one on product mix—got car makers into trouble a few years ago. Model schedules in 1970 were based on the previous year's experience—with small economy lines allotted only a relatively small portion of the overall market. Unfortunately, this assumption proved to be unwarranted as consumers became more and more price-conscious and began "trading down" to smaller and cheaper cars.

This caught some manufacturers, who didn't have enough of these lower-priced, less frilly lines in stock, by surprise. Detroit lost two ways: They had to surrender part of their market to foreigners, who were better able to meet the demand at that time. Also, they had to sell off their excess inventories of larger cars at discount prices. Again, nobody could have predicted this shift—much of it was psychological. Nevertheless, because such changes weren't monitored on a weekly or monthly basis, auto profits were adversely affected.

More and more firms are dealing with this need for greater production flexibility. The only answer is more periodic checkups of basic assumptions. The extra cost of these checkups is small compared with the loss that could result from sticking to rigid schedules.

Projection Pitfalls

Lack of flexibility to meet changing assumptions is only one of the many things that cause forecasts to go off the beam. Any time an analyst gazes into his crystal ball, he is in a sense sticking his neck out. With so many variables affecting any commodity, some kind of error is, of course, unavoidable. The various sources of error will be spelled out in detail in the next chapter. The emphasis here, however, is on showing how some of these errors can be minimized by avoiding or sidestepping the more common pitfalls, discussed below.

1. *Ignoring seemingly unimportant factors.* Often a particular variable may seem to have little influence on the item being projected. This may lead to the junking of this "insignificant" variable on the questionable grounds that it will continue to play only a minor role. But this isn't always warranted. Indeed, in some cases the insignificant variable may be the key variable in some later period—for a variety of reasons.

For example, in the early 1960s the price level of copper seemed to play a minor role in determining the metal's consumption level. A rise or decline of 1 or 2 cents had little or no significant effect on the tonnage consumed. But by the late 1960s a worldwide price spiral resulted in a near doubling of the price—from near 35¢/lb to over 60¢/lb. Suddenly many users who heretofore were unruffled by small price changes decided that a doubling in price was too much—and they made a concerted effort to shift to aluminum and other substitute materials. As a result, copper consumption for some applications—particularly where conductivity of the metal played a key role—began to fall off even though the business in general was still in a strong uptrend.

The moral is clear: A shift in the magnitude of a determining variable can make for a sharp difference in its overall effect. In the above case, prices which were insignificant over narrow ranges became a significant consumption determinant when the price range widened.

At other times a seemingly unrelated technological change can affect a projection over the longer pull. The introduction of the BOF steel furnace and its eventual effect on scrap consumption is a case in point (see Chapter 13).

2. Extrapolating beyond the range of experience.   The example of the price level of copper, discussed above, can also be regarded as one of a series of problems where the analyst is extrapolating beyond the range of his experience. In this case he didn't know what effect sharp price gyrations would have on copper consumption, but he assumed they would have as small effect as the limited price changes that had heretofore hit the red metal. As it turned out, this assumption was unwarranted.

Similar question marks crop up whenever an analyst assumes that a relation which existed in the past will continue to exist in the future— even though the magnitude of the numbers in the future will be considerably different. As noted in Chapter 15, there is little reason to assume that a nation will continue to spend and save the same portions of its income as it becomes more affluent. Maybe it will, and maybe it won't. But there is no a priori evidence either way. It is conceivable, for example, that as people become more affluent, they will choose to save a bigger percentage of their income.

Another "beyond the range of experience" error is assuming that what is true for one particular geographic area will be true for another. This time consider the relation between steel consumption and the GNP. We know that a certain number of tons may be associated with a given per capita GNP here in the United States. It is not always safe to assume the same for another nation when it, too, reaches the same GNP level.

For one thing, the life-style of the other country may be different. Then, too, the technological ground rules may have changed by the time the assumed level of income is finally reached.

3. Unrealistic extrapolation.   The fact that a certain product is growing at a given annual rate is no reason to assume it will continue to advance at that pace. Aside from the historic evidence of the "fast-then-slow" pattern in the life cycles of most product lines, there is the unrealistic element to consider. Sometimes the indiscriminate use of statistical variables in an equation can lead to nonsensical results. Take the example cited in the next chapter concerning an uncritical projection that would have led to a forecast calling for 10 TV sets per capita in the United States less than a decade after TV was first introduced. It points up the ever-present danger of blind extrapolation of past relationships and the need to temper forecasts with a dash of common sense and business savvy.

4. Unrealistic accuracy expectations. As also pointed out in the following chapter, accuracy is a function of time. The further one moves out into the future, the less certain the forecast is. This too, may seem nothing more than good common sense. But all too often statisticians, wrapped up in theory and mathematical formulas, tend to see their results as some kind of divine revelation which is correct to the last decimal place. Thus they will assign the same degree of reliability to the next month's forecast as to the next year's forecast—defying all the laws of common sense and statistical probability.

In this connection it should also be pointed out that no forecast can be made with 100 percent certainty, so it behooves the analyst to present his projections with a range of probable errors. Indeed, some reputable statisticians prefer to give a whole series of ranges. Thus it may be 50 percent certain that sales of widgets will be in the 3.5- to 3.7-million-unit range; it may be 80 percent certain that they will lie in the 3.4- to 3.8-million-unit range, etc.

5. Nonprojectability. It is not always true that everything can be projected. There are numerous instances, for example, where a survey result cannot be equated with marketing action. In short, we cannot always draw generalizations which will help us make the right decision. If we can't, then doing a survey is basically a waste of time and money—with the ensuing analysis little more than an exercise in rationalization.

Some analysts dodge the problem by presenting survey results without comment—and then let those who will generalize thereon. They escape making any judgment, and if someone should just happen to misinterpret the results, it is not their fault—these statisticians never said the survey was or was not projectable.

6. Incorrect accent. There is seldom one forecast—say, demand for widgets next year—that is perfect for each and every company purpose. Normally one would expect to see such a projection in terms of specific numbers or dollar volumes. But this isn't necessarily the most useful for all purposes. Product planning, for example, might involve a problem of whether or not to install another complete production line. The group's calculations might show that such a line would be needed if demand were to rise by more than 15 percent. For this need, then, a much more useful forecast would be one indicating the probability that demand will rise by more than 15 percent. This, in turn, may well require a different analytic approach from the previously noted specific number and value projection.

In a sense, before choosing a forecast the analyst must ask to what use the information will be put. In the above illustration he should ask, Are specific numerical forecasts required, or is the crucial problem the determination of probabilities that limits or thresholds will or will not be reached?

7. Projections based on averages. Every once in awhile serious mistakes can be made when a substantial dispersion or scattering exists. Assume that a firm wants to explore the market for some type of industrial bearings. In looking at a particular industry grouping, the analyst conducts a small sample and finds that the average plant in the grouping uses 19,880 bearings. He might then conclude that this is a lucrative market and go ahead with market expansion plans.

But he could be wrong. If 50 plants were sampled and 47 of them used 2,000 bearings, while the other three used, say 300,000, then using the small sample average of 19,880 is fraught with danger. Perhaps the group being sampled wasn't homogeneous. Perhaps the better plan would be to look at the three plants as a separate subgroup. In any event, using the small 50-plant survey average of 19,880 and projecting this to every plant in that group could lead to serious error in estimating the total market potential.

Composition

Errors stemming from a shift in mix are probably among the most common. Whenever a forecast is made, there is usually the implicit assumption that the composition or mix involved will remain unchanged. But again this need not be and often is not the case. A rundown on some of the more common types of composition pitfalls follows.

1. *High-value–low-value shift.* A few years ago most leading overseas steel-exporting nations signed an agreement with the United States that they would cut their steel shipments to this country sharply. Specifically, they said that they would drop tonnage from the 18-million-ton level of 1968 to only 14 million tons in 1969. This, it was assumed, would slow down the attrition in the United States balance of payments that had been plaguing this nation at that time.

At the end of 1969, when the figures were tallied, a curious paradox appeared. Shipments into the United States had indeed been cut by the agreed-upon 4 million tons. But the positive effect on the balance of payments was negligible. What had happened was that overseas sellers had stayed within the volume limitation but had shifted over into more expensive product lines in order to keep their earnings from falling.

Thus in 1969 the dollar value of United States imports (heavily weighted with expensive tool and stainless steels) came to $1.7 billion. That was down only 12 percent, as contrasted to the nearly double 22 percent decline in the physical volume of steel imports.

The value-volume pitfall is ever present—even where there is no premeditated effort to circumvent a ruling or goal. The dynamic forces within our economy make it exceedingly unlikely that the same product mix will be repeated year after year. That's why value and volume forecasts should be analyzed separately. Looking at both series separately also helps isolate the effect of inflation and changing price levels—a subject that was discussed more fully in Chapter 11.

2. *Composition and substitution.* The classical example here involves the evaluation of the consumer price index. Other things being equal, a sharp rise in the price of one good will result in a shift to a substitutable good which has not gone up in price or which has gone up only fractionally. Since a price index has fixed weights, this shifting doesn't show up in the index level, and thus the index tends to have a measurable upward bias. A fuller discussion of this kind of bias in index numbers may be found in Chapter 6.

The rise in the consumer price index is sometimes thought to be more than it really is for another reason. The consumer—as his standard of living increases—always tends to upgrade his purchases. But because he is relatively unsophisticated, he thinks that prices are going up. Take, for example, the storekeeper who complains about inflation, citing the doubling of his electric bill over the past year. When it is pointed out to him that utility rates haven't changed over this period, his classic reply is, "No, but I just put in central air conditioning."

The problem here, of course, is that storekeeper is confusing his cost of living with a price index. The sharp rise in utility costs to which he refers pertains to his own decision to raise his own cost of living—not to any noticeable rise in the price level. But the government doesn't have a true cost-of-living index; it has a price index. And the two can vary substantially—particularly when standards of living are changing rapidly.

3. Composition and service prices. Over the past decade or two, our economy has become increasingly service-oriented. Indeed, today consumers spend more on services than they do on durable or nondurable goods. Since services are prone to inflation, it follows that it is becoming increasingly more difficult to contain inflation than it was in the 1940s and 1950s. In other words, if we pursued the same anti-inflationary policy 20 years ago and now, the price rise today would probably be considerably bigger because so many more services are now included in the average.

4. Composition and comparison. An illustration culled from the social science area can best highlight this problem. In recent years the statistical disparity between the incomes of Negroes and whites has been attributed almost entirely to prejudice. But a look at the facts would indicate otherwise. A large part of the difference in median family income is due to the fact that a disproportionate number of blacks live in the rural South, where incomes are lower for everybody. Indeed, take this into consideration, and the statistical disparity between the incomes of blacks and whites is very nearly halved.

In the business arena this is akin to comparing wage rates in one widget factory with those in another—both organized by the same union. The fact that there are more highly skilled people in the first factory than in the second is no reason to state that the workers in the second factory are being discriminated against. If the composition of the workers were analyzed, it would be clearly seen that the wage rates for comparable skills are the same.

5. Fallacy of composition. This is a classic case cited in economic literature which would lead one to believe that what is good (or bad) for the individual is good (or bad) for the society as a whole. It all hinges on the often erroneous equating of individual and collective action.

It can best be seen in terms of an economy in which some people are threatened with layoffs. Under such conditions, it would make sense for an individual worker to pull in his horns—buy less and save more. But if every worker were to follow this strategy, it could well hasten the feared result: job layoffs. Specifically, less buying would mean inventory glut, production cutbacks, and eventually job layoffs. In short, what made sense to the individual worker would be disastrous if followed by each and every worker. This, incidentally, is why when a recession threatens, the government tries to play the fear down. Only by rebuilding confidence can the government prevent the fear from becoming fact.

Business is also subject to the fallacy of composition. Suppose one manufacturer in an industry decides to increase production. He can reasonably expect that the increase will bring him more money if demand is strong. But suppose that all manufacturers in the industry decide to expand. Will they all get increases in profits? Not necessarily. It is quite possible that increasing supplies will push prices down, reducing profits for all concerned. Clearly, what is good for one manufacturer is not necessarily good for all manufacturers acting in unison.

6. Composition and appropriate yardsticks. With the real world so varied, it is often difficult to choose the best measure. In the case of price indices, there are literally thousands of them published by government and private sources. And if one wants to deflate a value figure, one has to be sure that the composition of the price index closely resembles the composition of the value aggregate to be deflated. If it doesn't, the resulting volume or deflated figure can yield questionable if not meaningless results.

In the same vein, a company that wanted to compare its own price with that of its industry would have to make sure that the product mix of the industry closely approximated its own. Similarly, if the firm wanted to make inventory, order, production, or sales comparisons, the comparability of product mix would be crucial. In some cases, it is possible to rejigger a reported figure (possibly with the aid of the issuing agency) to bring it more in line. To be sure, this will often involve considerable time and money. But to take the easy way out and use an inappropriate yardstick is no answer either.

Sometimes a little exploring will turn up a better-than-expected yardstick that is already available. Some wage contracts, for example, are escalated in line with the national consumer price index. But it is much more appropriate in many cases to use an available metropolitan-area consumer price index—particularly when local contracts are involved. A surprising number of labor analysts who should know better are often not even aware that these pertinent price breakdowns exist.

Statistical Detail

The complexity of products and markets makes it increasingly difficult to make one figure or one total apply to all operations. Delving more deeply into the makeup of production, sales, and other key business parameters has become a must for intelligent decision making. The big question, however, is, How much detail?

Too much detail can be as self-defeating as too little—weighing down decision makers in a morass of insignificant minutiae as well as adding to costs. Here's a rundown of ways of breaking down business statistics, plus a few general rules on how far to go in each.

1. Product detail. For a medium-sized firm making, say, two or three basic chemical lines, there is little need to go into much detail because of the homogeneity of the product lines. On the other hand, an automobile manufacturer with more than 25 models to contend with has to go into great detail because sales depend to a large extent on the availability of the specific model when it is demanded. Similarly, a large firm with access to a computer can make much more detailed product breakdowns than a smaller firm where every product breakdown necessitates expensive and tedious clerical work.

2. Market breakdowns. Market breakdowns can be approached from several different vantage points—by end use, by class of customer, and by type of distribution or trade channel. The typical end-use approach might involve a basic raw-material or component-parts producer who will break out his forecast on the basis of anticipated sales to industries consuming his products. Such a split-up is useful because very often the supplier has advance knowledge on the outlook of the user industries—and hence has some pretty good ideas on how much he can sell to each of these industries.

Analysis of past sales to specific industries also points up where a supplier is making inroads against competing products and where he is not—and thus can give hints on his long-term sales trends. Aluminum producers, for example, point to a 67 percent gain for aluminum consumption in cars (from 45 pounds per unit to 75 pounds per unit) in the decade ending in 1965. On the basis of this trend, aluminum producers in the latter year confidently predicted a 125-pound-per-car consumption figure (another 67 percent gain) by the end of the next decade (1975).

The problem, however, is that not all firms can use the end-use approach because of statistical information gaps. Thus, not all firms can emulate the steel and aluminum companies that code incoming orders by end use (auto, appliance, transportation, construction, etc.) and then use this information as a clue to their future sales volume to each of these industries.

The common complaints by those who would like to use this approach, but can't, are (1) "Our sales are too broadly scattered for this purpose" and (2) "We make most of our sales through dealers, who are not about to set up additional work for themselves—namely, tracing and tabulating the type of customer who buys the product."

Market-type breakdowns, however, need not be limited to the end-use approach. Detailed analysis is also often performed by splitting up overall sales by class of customer and trade channel. As for the class-of-customer approach, if a company sells to both consumer and industrial customers, a breakdown of the sales forecast to each might be useful, inasmuch as different variables affect the sales trend for each.

Similarly, a breakdown by trade channel is often important when a firm sells some of its products directly through its own sales force and the rest indirectly via distributors, dealers, brokers, etc. Differing trends in each, for example, can suggest areas of strength and weakness and can point up areas where the distribution setup needs change.

3. Customer breakdowns.   In some heavy-equipment lines, where only a few customers make up the bulk of the sales, it is important for the manufacturer to forecast sales on a customer-by-customer basis, since any individual purchaser can make or break the sales projection. Makers of heavy electric equipment often proceed along these lines, asking the relatively few utility companies what their capital equipment plans are over the next few years.

This same approach could conceivably apply to a consumer goods manufacturer who has most of his production contracted on a private-brand basis to a few large chains or mail-order houses. Here, again, a few individual customers call the tune on the overall outlook.

4. Geographic breakdowns.   Geographic breakdowns are quite useful because most national firms set up their sales force on a regional basis. This automatically calls for territory-by-territory sales intelligence—both as guidance to regional management and for setting individual salesman quotas. Moreover, such breakdowns suggest areas where more or less effort may be needed to maximize sales potential.

Changing Share

Too often the criterion of success is how fast and to what extent a firm can increase its share of the market. A few moments' thought will indicate that this is not always the best measure of a firm's progress. Under some conditions a maintained share, or even a declining share, might not be as catastrophic as share percentage alone would seem to indicate—particularly if growth was attracting many new suppliers into the market. Under these conditions absolute sales would still be increasing at a fast clip, and it would be the height of fantasy to assume that such prosperity conditions would not attract competitors.

On the other side of the coin, a rising share in a declining market can often be misleadingly optimistic. One marketer tells of a declining chemical (replaced by a cheaper and more efficient substitute) which was losing sales at the rate of over 30 percent per year as buyers gradually adjusted their own manufacturing process to accommodate the new substitute.

This executive added that other suppliers were dropping out of the market and that as a result, his company's share of the declining market was increasing rapidly. But he figured out that just to maintain a break-even point as far as volume of operations was concerned, the company would have to operate at about 150 percent of capacity within just a few years.

An increasing share is also not always consistent with improving profits. If the price is declining, what good does it do to expand share when the firm is producing at a loss or at very little profit? The old anecdote comes to mind about the outfit that tried to increase its sales at the expense of competitors by cutting prices below actual costs. When an executive was asked how the firm could survive while losing $5 on each sale, his answer was, "Volume, my boy, volume."

The story is funny because it points up the incongruity of setting one's goal exclusively in terms of quantity share. The point, of course, is that growth in volume alone is a meaningless measure. It takes on importance as one of several ingredients in the profit outlook—which, of course, is the ultimate concern of every business manager.

The tendency of some to assume that an increasing supply of a good or service will result in a diminishing share for each individual supplier is also open to serious question. For example, taxi drivers for years have opposed an increase in the number of metered cabs on the grounds that this would lead to a drop in business for each individual driver. This may or may not be so. Certainly there is no a priori reason for concluding that it would automatically reduce per-cab business.

If a greater number of taxis were made available, this could conceivably increase rather than decrease the total market. This would be the case, for example, if many people who would normally use this kind of transportation have not been doing so because they can't find an empty cab when they want one. The point is that an increase in supply (the number of vehicles) does not necessarily mean that the share of work per supplier (per vehicle) must decrease.

FIFTEEN

A Question of Accuracy

In its triumphal march through the minds of laymen and apparently many businessmen, statistics has taken on an aura of precision and certainty that it doesn't really possess. Complete, 100 percent accuracy is the impossible dream—a goal to shoot for, but always just beyond reach.

Of course, for many purposes forecasts need not be all that precise. As most businessmen have discovered, projections that turn out to be even approximately correct can prove enormously useful both in making day-to-day decisions and in planning long-run growth strategy.

It all boils down to payoff. Every dollar spent on statistics represents a cost and hence a reduction in profits. Thus the only justification for a statistical outlay is that the increased accuracy stemming from the use of such data will bring back the dollar, plus a couple of "friends" along with it—preferably, it's hoped, in a reasonably short period of time.

Increasing statistical sophistication, in general, has permitted the attainment of such a goal. But it wasn't always so. In the sphere of general business forecasting, for example, it has been said that 20 years ago the one sure way to gain fame was to take the average of all forecasts made that year—and then come out with precisely the opposite projection.

To be sure, this is an exaggerated statement—even for those early, developing days of statistical analysis. In any case, today's forecasts are much better—indeed, they are surprisingly accurate in view of the many complex forces confronted in the real world. This real-world aspect, meantime, raises still another problem: the best approach to statistical forecasting. Specifically, how much should an analyst rely on mathematical techniques, and how much on subjective judgment?

In most cases, a combination of both approaches is usually best. One market analyst puts it this way: "Unquestionably, there is an urgent need to take advantage of insights that cannot be caught in the net of quantitative statistical techniques. Thus models should be designed so that they can easily incorporate judgmental elements. But this should be done without violating the integrity of the mathematical technique."

In the final analysis, statistical projection is a combination of art and science. This lack of pinpoint mathematical accuracy, in turn, suggests an approach that might save statisticians considerable embarrassment: the presenting of results in terms of probability, much as weather forecasts are made. To this can be added the presentation of a range wherein the actual results are likely to fall.

Calculation of such error ranges should factor in both sampling errors (see page 336) and nonsampling sources of possible variation. Thus, there are errors in basic data, errors in calculations, unforeseen events (strikes, political upheavals, etc.), and changes in the underlying assumptions. Too often analysts tend to ignore these latter sources of error because they are too difficult to quantify. But pretending they don't exist is hardly the answer—nor is quantifying them with wild guesses, which is equally bad. A substantive body of empirical evidence is usually available to guide the assessor in his quantification of all possible sources of error.

Finally, before going into a more detailed listing of the various types of error, it might be well to point out that there are also varying perspectives for viewing accuracy. In a sense, accuracy is a relatively complex term. The layman usually thinks of accuracy in terms of how close projected results are to actual results. This element of accuracy, for want of a better expression, can be termed the *variation* factor. The smaller the variation, the more accurate the forecast. As will be shown later, variation in itself can be viewed from several different perspectives.

But accuracy can be a lot more than just variation—no matter how defined. It should also take into account whether the forecast is a lucky one or is built on solid, logical foundations. This can be looked upon as a *validity* factor. For example, it is entirely possible for an aggregate projection to be correct because of a fortuitous canceling out of large errors in components, but this is hardly the way to engender confidence in a forecasting approach.

Also included under validity is the use of consistent methodology and consistent data. The researcher who changes his basic data source or technique from period to period merely to reduce variation should automatically be suspect. A valid approach should yield good results today, tomorrow, and the day after tomorrow.

Sensitivity is still a third factor in evaluating accuracy. This boils down essentially to the question of whether the techniques used pinpoint the underlying trends.

EXAMPLE: Take two sales forecasting approaches. The first may yield a forecast that is only 2 percent off, with actual sales going up 1 percent, while the projection called for a 1 percent decline. On the other hand, the second forecasting approach may also yield a 2 percent error, but one traceable to the fact that the projection called for a 3 percent rise in sales, in contrast to the 1 percent rise that actually materialized.

The two approaches have the same degree of variation accuracy to the extent that the variation of observed from projected data is exactly the same. But the second approach is the more reliable or more sensitive one, in the sense that it has correctly indicated the direction of sales. This is an important consideration—and for most purposes is even more crucial for basic policy decisions.

Input and Mechanical Errors

More often than not, the most important source of error can be traced to a simple cause—such as inaccurate or incomplete raw data, arithmetic errors involving their manipulation, or sometimes even the use of some inappropriate procedure. At other times the length and scope can affect accuracy; other things being equal, the narrower the product line covered and the further out into time a projection is made, the greater the chance of errors and inaccuracies. Each of these aspects will be examined below.

Raw data. There is a general tendency to assume that the given raw data are completely reliable—and there is often little attempt to delve into how the information was collected and tabulated. Such assumptions are often unwarranted. For example, surveys of consumer spending may depend on how well the consumer remembers what he bought and how much he paid and on whether he will report all his bad purchases.

Similarly, surveys of unemployment include some people who are not seriously hunting for jobs, while leaving out the women who will be job hunting as soon as their husbands or fathers are out of work. In general, many statisticians believe that official unemployment tabulations greatly underestimate the seriousness of the problem.

One senator, critical of the entire measure, details his objections: "They fail to count those who have finally given up after years of frustration and dropped out of the labor market altogether. And they fail to count those who are so poor and so transient that they have never appeared on a census list and can never even be polled by the government."

Another criticism leveled at this measure of joblessness is that it utilizes faulty adjustment for seasonal variations. Many government people go along with this last point—and say that's why this measure often tends to be so erratic. The Bureau of Labor Statistics—the issuing agency—also admits to the possibility of error. Because of possible sampling error, for example, they can promise only that in 9 out of 10 cases, the month-to-month changes in the unemployment rate won't be off by more than 0.12 of a percentage point in either direction. This means that the rate usually has to move at least 0.2 of a percentage point before the change can be considered statistically significant. In today's labor force such a percentage could involve as many as 160,000 people.

The same limitations can be leveled at many other key series. Take capital spending—particularly the discrepancy between planned and actual performance. In some cases the variation has been as much as 5 percent of the total outlay envisioned.

Although survey organizations attempt to adjust results for systematic biases, anticipated amounts of spending have consistently exceeded actual outlays in recent years. That's not to suggest that these surveys do not provide valuable information. Rather, the point is that anticipatory data, no matter how carefully collected, are usually open to a considerable margin of error. Indeed, other things being equal, anticipatory data (on both the consumer and the business levels) are usually on a lot shakier ground than any other type of basic statistical information.

There are ways, however, for reducing the amount of error and making the expectation forecasts more useful. Thus a few years back the Commerce Department, in conjunction with the National Bureau of Economic Research, developed a methodology on consumer anticipation surveys whereby respondents were asked about probable purchases of automobiles and houses on a continuous scale of probabilities from 100 (absolutely certain) through 0 (absolutely no chance) and about the price they expected to pay if they bought. And it has been paying off. Since the initiation of this new procedure, there has been a marked improvement in the accuracy of consumer surveys.

Sampling procedures can also be improved upon. For example, in the improved consumer survey noted above, data are obtained every three months from a random sample of approximately fifteen thousand households. The full sample comprises six independent subsamples (rotation groups) of about twenty-five hundred households each. About five-sixths of the sample households are identical in consecutive surveys, a fact which tends to minimize sampling error in the measurement of changes.

A further improvement, added to this survey by the Commercial Credit Co., is the translation of responses into estimated dollar sales. Specifically, projections for household durable outlays and single-family housing outlays use Census Bureau data in conjunction with economic and financial variables like income, unemployment rate, etc. Projections for unit sales of automobiles use Census Bureau data on car-buying expectations; the index of consumer sentiment, developed by the Survey Research Center at the University of Michigan; and other financial and economic data.

Coming back to the basic raw-data problem, statistics arrived at via the residual approach are also open to more than the normal amount of error. Thus in national income analysis, savings is what is left after private consumption is subtracted from disposable personal income. Since both of these are very large compared with the amount of savings, small errors in income and spending estimates can lead to large percentage errors in the estimate of savings.

Unfortunately, warnings of inaccuracy or possible variation are not always included when this and other forms of raw data are presented. Users of statistics are not always sufficiently warned of the limitations of the material and the possibility of misinterpretation. And when there is such a warning, it is received only by the analyst. More often than not, he will forget—sometimes on purpose—to include it when forwarding his finished report to management.

Another data problem arises from the fact that in many cases the information is compiled by business and government—not primarily to facilitate economic analysis but to support management or regulatory controls. Hence it may not suit the needs of the analyst, and from a forecasting viewpoint it may embody conceptual weaknesses.

To sum up, accuracy problems arise not only with government statistics but with all data, whether of a business, economic, or a scientific nature. Possibly the soundest approach is to assume that all data contain some error and that it is therefore necessary to determine the cause of the inaccuracy, its seriousness, and its ultimate effect on the analysis being performed.

Arithmetic errors. Even in the computer age, the possibility of a reversed number or a wrong keypunch cannot be entirely ruled out. There have been instances, for example, when a regular monthly government release has been delayed for more than two weeks because the computer reading suggested a far more drastic change than seemed warranted. Only when the basic data were checked and rechecked did the error become apparent.

Every once in awhile these errors manage to slip through. And only when a detailed checkup is made sometime later are they discovered. Thus many of the monthly revisions in government statistics are due not so much to the receipt of more up-to-date information as to the discovery of past computational errors. At one point a few years ago, retail sales were raised 2 percent three months after the fact, when it was discovered that sales from a relatively small metropolitan area had been inadvertently dropped from the tabulation.

The chance of such inaccuracies, of course, multiplies as the processing becomes increasingly complex. Thus a recent input-output report, originally scheduled for late 1968, wasn't issued until a year later. Each of several postponements was blamed on "computer difficulties"—a euphemism for "the answers on the computer make no sense at all."

Data revisions. Aside from revisions due to errors, many changes are due to the receipt of new benchmark data. For the issuing agency this usually presents few problems. The figures are just reissued for the pertinent periods. But for users of these statistics, it can be a source of considerable trouble—interfering with comparisons (see Chapter 13) and often necessitating major recalculation of past relationships.

There's no way out of this dilemma—except to be aware that such revisions are being made and to be ready to plug them into your own equations as soon as they become available. To ignore such major revisions can be dangerous, for sometimes changes can be quite significant. Thus in predicting the level of the GNP over the following year, one expert found that revisions made in the estimation of the previous year's GNP were the major factor in the total forecast error (see page 348).

Cost. Budgetary considerations also limit the accuracy of statistical information. Cash outlays obviously must rise with the degree of accuracy achieved. Thus companies must weigh the benefits of more precision against the additional cost incurred. Clearly, these cost restrictions have commanded less accuracy than would be professionally feasible.

The role of coverage. Accuracy is a function of degree of coverage. The more items and the more areas covered, the smaller the chance of error in the overall prediction. A moment's thought will indicate why. When many items or areas are covered, errors tend to cancel out. Statisticians refer to this phenomenon as the "law of large numbers."

Take a hypothetical case in which a company makes two products, A and B, in equal volume. The firm feels that both should gain about 5 percent and hence predicts a 5 percent overall advance in sales for the coming year. This 5 percent prediction could come out right on the nose, even though the assumption of a 5 percent gain for each item was way off. Conceivably, one product could show a 10 percent gain, and the other none at all. There are a myriad of other percentage combinations that also could yield the same 5 percent overall gain.

Considering that the average firm makes hundreds or even thousands of products, it's easy to see how literally millions of different combinations can bring about the correct overall result even though the figures for the individual products are way off.

The same coverage factor helps explain why an industry forecast is usually more accurate than a company projection. The industry estimate doesn't have to contend with the individual variation (both up and down) stemming from competition and other factors which influence individual company share. These volatile factors cancel out, or wash out, in the industry forecast.

Accuracy and length. Just as error is a function of coverage, it can also be thought of as a function of time. Generally speaking, the further one goes out into the future, the greater the chance of error. Put another way, accuracy is a decreasing function of the span of the forecast.

Unfortunately, little can be done about this because of the nature of the forecasting approach. Both data availability and possible changes in the forecasting relationship can be responsible for error proliferation over time.

On the availability front, the clues or the variables that, for example, explain sales are already pretty much "in" for the next few months. Thus a department store owner has a pretty good idea today of what disposable income will be three months from now, and hence he can apply his time-tested sales/income relationship to predict sales over the short run.

If he has to predict sales one year from now, he is faced with a more difficult problem. For his estimate of income 12 months from now isn't likely to be anywhere near as accurate as his estimate of income two or three months from now. The clues to future income are just not as good or reliable for the longer pull.

There's an additional source of error in long-run projections. Even if the data input, by some stroke of luck, is 100 percent accurate, there is still no assurance that the basic relationship between, say, sales and the variables that explain sales will hold, for we live in a dynamic society, and new forces or influences are always impinging upon established relationships.

A good example of this occurred a few years ago in the metal field. For years, sales of copper transmission wire were closely correlated with the electrical utility estimates of future capital outlays. In the later 1950s and early 1960s aluminum began to take over from copper in this area. Anyone who used the old wire–capital outlay relationship of 1955 to predict copper wire sales today would be overshooting the mark by a fantastic amount.

Analytic Errors

Other types of inaccuracy stem not so much from mechanical mistakes, errors, and shortcomings as from the application of the wrong type of procedure. These are difficult to document only because there are so many opportunities for erring. For example, a poor choice of an explaining variable in a correlation could make for underestimation or overestimation of, say, inventory or capital spending requirements. Similarly, a poor deflator may lead to underestimation or overestimation or real volume of sales—past, present, and future.

This choice-of-variable problem can be solved only with intimate knowledge of the industry being studied, so no further general remarks can be of any value. On the other hand, a brief discussion of some of the other "families" of analytic pitfalls may be in order. These would include the following.

Short base period.   Since forecasting is based primarily on historical relationships, the past record must be of sufficient length to assure validity. One expert, while he admits that no a priori or empirical answers to the question are available, points out that it would be "statistically unreal or unreasonable to use one or two time units such as years, months, or days for the purpose of projecting five or more equivalent time units." This expert goes on to say that "a rough rule of thumb employed by some analysts is that the length of the base period should be at least equal to the term of the projection period." Only then can the results meet the rigid mathematical criteria for a forecast that is both valid and useful.

Extrapolation beyond the range of experience.   This kind of projection error—discussed also in the preceding chapter—tends to crop up most often in correlation problems. It can best be described by a simple hypothetical example.

Assume that historical data reveal an excellent relationship between small-appliance sales and family income. Specifically, sales tend to rise 1 percent, say, for each $100 rise in annual income. Furthermore, the relationship has been derived from past experience in which income has varied from $5,000 per household to $7,000 per household.

So far so good. But what happens if income spurts to $8,000? Does the same relationship hold? One can't be absolutely sure. It is a little risky to assume that the same 1 percent rise per $100 of income will prevail when income goes much above the $7,000 mark.

Trend perception.   The temptation to project short-run developments into long-run trends can often lead to error, or at least to a difference of opinion. For example, some see the increased purchase of services (relative to goods) in recent years as a trend acceleration. Others, pointing to a similar development in the 1920s, say "no." They feel that reviewing over a sufficiently long period of time reveals no trend acceleration, but only two bumps on an otherwise smooth trend.

The same difficulty has evolved around productivity estimates. In the early 1960s manufacturing productivity was increasing at 4 percent or more per annum, compared with the 2.5 to 3 percent rate of the previous decade. Some analysts claimed that this was a new trend. Others, pointing to previous spurts which later fizzled out, said "no." Who was right? It will probably take another decade to get the answer.

Estimates on estimates. This type of projection is also generally associated with correlation problems. Here the problem involves the need to estimate an independent variable before making a final projection for the dependent variable. Thus in predicting next year's sales, one would have to make an educated guess about next year's income (the independent variable). This means that the normal amount of error in projecting sales must then be multiplied by the possibility of error in income (the sales-determining variable). With two areas of possible error, the chance of variation is substantially expanded.

As a rule of thumb, if it proves almost as difficult to estimate the independent variable as to estimate the dependent variable, then regression analysis may not be the best way to tackle the problem. Similarly, estimates based on inaccurate basic data (discussed above) will again almost always multiply the error.

These is also a compounding effect when the original figure and the final result have been estimated incorrectly. For example, a 10 percent overestimate on an original estimate that is already 10 percent above the mark could conceivably yield an error as large as 21 percent ($110\% \times 110\%$).

Estimating lags. One major source of statistical error is the inability to estimate lead-lag relationships with precision. Thus while it is clear that a sizable cut in income will curtail sales of big-ticket appliances, it is hard to say whether the reduction will come one, two, or three months after income has been cut. The problem here is that techniques are still not sensitive enough to pinpoint the timing relationship.

This is particularly irksome in the use of monetary and fiscal policy to control business growth. Take the behavior in the period 1968 to 1969. First, federal taxes were raised in mid-1968 to cool off a runaway boom. But the public promptly went on a spending spree; consumer spending did not slow down until several months later.

Money, meantime, was also tightened. But except for the homebuilding industry—always the first victim of tight money—business activity did not begin to reflect the squeeze until more than three-quarters of the way through 1969.

To sum up, not only do the real consequences of policy changes lag, but the lag is also variable—sometimes short and sometimes dragged out over time. Those who must determine policy are dealing with tools that they know will work. But they cannot be sure either exactly when the tools will work or how extreme their effects will be in any given circumstance. In short, the application of statistical techniques to economic problems (econometrics) is still an inexact science.

Forecast feedback. This type of error, usually associated with projections, generally stems from not taking into account corrective measures that management may take to bring the forecast more in line with its basic targets. If, for example, profit projections are under desired levels, management may take new steps (not assumed in the forecast) to beef up the earnings level. Some statisticians refer to this as the *Heisenberg effect*—after a famous physicist who stated that the process of measuring a phenomenon often changes the parameters of that phenomenon.

Care must be taken not to blame the analyst for variation stemming from this kind of feedback pressure. Indeed, if anything, an accurate forecast of low profits may have performed a needed service—alerting management to the fact that current policies were not consistent with the firm's basic goals. On the other hand, some of this error can be reduced if the analyst, when coming up with his first tentative forecast, checks with management to see whether any change in the underlying assumptions may be necessary. If so, he can then change his forecast— and at the same time note that the more favorable forecast is premised on additional company action, which is then spelled out.

Noneconomic Factors

A forecaster can hardly be expected to predict such an unexpected happening as a wildcat strike or an international crisis. Yet these outside shocks can significantly influence dollar and volume data. For evidence, one has only to look at the scare-buying sales bulge of the Korean War (1950 to 1951) or the sudden drop in auto sales in late 1964, when union locals failed to go along with an industry wage settlement.

The wide variety of such noneconomic, one-shot disturbances was discussed in great detail in Chapter 11. They are brought up here again only to stress that they can introduce substantial errors in what otherwise might be a statistically airtight analysis and projection.

Also, as pointed out at that time, there are ways of dealing with or at least minimizing errors from these unexpected sources. Thus in the case of, say, steel negotiations it's almost always a good idea to factor in heavy steel hedging about six months before the strike deadline. Similarly, in predicting beer sales during the summer, it might be possible to give a range of forecasts based on a possible range of summer temperatures. In any case, where forecasts are usually subject to such unpredictable influences, it is always a good idea to hedge them with a listing of the major assumptions involved.

Subjective Errors

In some cases errors are due not so much to the shortcomings of the statistics or techniques as to the shortcomings of the analyst. The facts and figures are all there, but for some reason or other they are evaluated in such a way as to increase rather than minimize the possibility of error. This is not to say that the subjective evaluation isn't useful and indeed necessary. Rather, the problem is that this subjective element is not being applied to achieve maximum benefit.

At other times subjective errors occur because the objective information needs refinement but the exact extent and direction of such refinement isn't clear. The evaluation of information received from a survey— or of a firm's competitive position in an industry—is more often than not open to question, and hence the possibility of mistakes. The most common types of subjective errors include the following.

Bias. Several kinds can be listed. First there is the so-called unintentional intentional bias. Here the analyst does not want to be misleading. But because he is so involved in proving his point, he finds it increasingly hard to remain objective. He may attempt to present accurate information, but because he is obsessed with his goal, he unintentionally ignores any data which might refute his basic hypothesis.

Second, there's the conservative bias. When forecasting, there is usually a tendency to be a bit on the conservative side—primarily because there is less risk involved in following this approach. This, too, must be considered a bias because the analyst is being a little less than totally honest when he presents his statistical findings. More will be said about this conservatism in the next section.

Other, more obvious types of bias problems occur in instances where there is a clear-cut aim to mislead or hoodwink the recipient of the statistics. Many of the examples given in Chapters 12, 13, and 14 illustrate the myriad of ways in which this deliberate or intentional bias can be effected.

To counter this intentional kind of bias it is often advisable to check on data sources. The competitive nature of business organizations may often prevent analysis from showing both sides of the picture. But for the analyst, this two-sided information is a must. An example from the field of politics should make this clear. Specifically, if a liberal-leaning person wanted objectivity, he would be foolish to read only analyses and polls of liberal-thinking organizations. The elimination of bias and the obtaining of a more rounded picture would clearly require reading some conservatively oriented reports. Conversely, conservatives seeking objectivity would do well to rely—at least partially—on information provided by liberal groups.

The point is that relying on data known to have a certain bias is an open invitation to trouble. And often the best way to correct this is with a little data with the opposite bias. This applies to business and economics as well as to politics. Thus before labeling a wage hike as too high or too low, it's a good idea to read both management's and labor's arguments.

In still other cases, bias can be traced to the basic collection rather than to any analytic shortcomings. In survey data, for example, respondents often answer without weighing all aspects of the question being asked. The result is misleading information—so misleading, in fact, that the government, before it publishes data on consumer and business buying intentions, generally makes a so-called bias correction.

Such corrections are based on historical experience with actual and reported data. Unfortunately, these bias corrections are less than perfect. Indeed, despite nearly a decade of experimentation, there still exists a wide gap between actual and reported results of government anticipation surveys.

Conservatism. When there is a choice between opting for little change and opting for much change, the former is almost always elected—even when the preponderance of evidence suggests the latter. Again harking back to capital equipment surveys, year after year (despite bias adjustments) actual results have tended to run ahead of forecast levels—far too often to be justified on the basis of pure chance.

The explanation in the case of capital equipment is relatively simple. Survey respondents, looking at past performance, chose to stick to time-tested relationships of capital spending to profit and capacity utilization as a basis for their projections. But as pointed out earlier, new forces have entered the capital equipment spending equation—notably the high cost of labor, obsolescence, and increasing competition. This, in turn, has meant more investment per dollar of profits or percentage of unutil-ized capacity. However, because nobody had been willing to go out on a limb and depart from the historical relationship, the totals have been consistently underestimated.

A similar underestimation shows up in projections of individual firms and industries. On the industry level, one major aluminum company (Alcan Ltd.) has admitted to a consistent underestimation of aluminum consumption (see Figure 15-1). To be sure, the error can be explained away in terms of bigger-than-expected inroads into both new and estab-lished markets. But this is hardly any justification. Aluminum, of course, should not be singled out for such criticism. This same conservative bias has occurred in industry after industry in recent years—and is prob-ably responsible for some of the material shortages we face today.

The basic reasons for this conservatism are the same in almost all instances: the inherent resistance of people to change—plus the fact

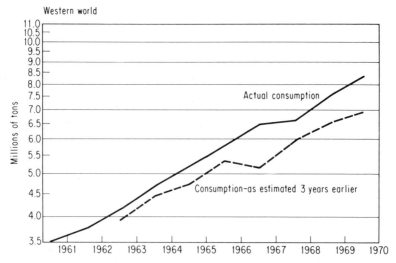

FIGURE 15-1 *Primary aluminum consumption—actual versus pro-jected results (logarithmic scale).*

that the system is set up so that it is better to err by projecting too little than by projecting too much.

A few additional words on this latter point: On an after-the-fact basis managers generally must explain why a particular item has not performed according to plan. This in itself is not undesirable. But in practice, companies generally tend to reward the manager whose performance exceeds his original forecast and to penalize the one whose performance falls below.

Middle-echelon managers have been quick to interpret the implications of this unofficial reward system. They react by opting for the conservative estimate for transmission to top-level management—well aware that they have a better-than-even chance of eventually exceeding this level.

Put another way, as long as overestimating is more threatening or less rewarding than underestimating, the analyst will almost always tend to protect himself by opting for the latter. This, then, suggests that companies should try to redress the balance by penalizing all inaccuracies (on both the plus and the minus side) with equal harshness.

Assumptions. Reaction to changing assumptions is also in many ways subjective. Almost any well-trained statistical analyst, for example, should be well aware of the various assumption pitfalls discussed in the preceding chapter. Nevertheless, in a surprising number of cases he may still choose to ignore such shifts. The basic question to be explored here is, What are the subjective forces that make a trained technician ignore these seemingly obvious changes?

On one level it may be done simply to avoid major recalculation—or maybe to avoid the necessity of sending revised forecasts up to the upper echelons. Some fear, for example, that frequent changes sent up to management may lead to charges or thoughts that "those people up in research just can't make up their minds." The answer here, of course, is that they shouldn't, if key ground rules keep changing.

Then, too, there is a tendency on the part of some to shrug off changes as inconsequential. This can be extremely dangerous. For example, when the 7 percent investment tax credit was eliminated several years ago, it was thought by some to be of little statistical consequence. Yet a checkup a year or so later revealed that this "inconsequential" factor had been enough to pull capital spending totals down some $1 billion below what they would have been had the tax credit remained on the books.

At other times, ignoring the shifting ground rules reflects basically sloppy work attitudes. Some researchers, having completed a job, promptly forget about it until the next such forecast is mandated from higher up. Wars can break out or strikes can occur, but these people choose to ignore them on the questionable grounds that they have already done their job. Happily, this doesn't occur too often. But when it does, it's high time for a change in attitudes or a change in personnel. Management is hiring a researcher to keep it apprised of an admittedly complex picture. And if the research staff doesn't provide this service, it is clearly not performing its function.

How, then, does one deal with the problem of changing assumptions? Aside from consistent checkups, it is generally a good idea to devote a separate section of each finished report to just this subject. The report should cover each and every assumption, indicating why it seemed most reasonable at that time. In addition, forecasters may want to give rough estimates of how the projection might change if a different set of assumptions were used.

Thus the auto companies in early 1964 gave two alternative estimates of yearly sales—the lower one on the assumption that a tax cut would not be passed, and the higher one on the assumption that it would. Since both eventualities were equally likely at that time, the companies were alerted to two possible different sales levels.

In addition to forcing the researcher to think about something more than pure numbers or pure technique, the emphasis on assumptions and how they affect the forecast ensures a documented record of the company's thinking at the time the outlook is being prepared.

The "yes" syndrome.   Here the problem involves the supplying of management with reports the statistician thinks management wants. In many cases this, too, stems from a basic misunderstanding of the statistical function. The function of the analyst is that of a fact finder not a policy maker—and thus it is incumbent upon him to present such facts in the most objective way possible.

In all fairness, however, it should be pointed out that the analyst is not always a free agent. A vice-president or a division head who is bent on launching a pet new project may persuade his research staff to come up with the "right" cost and profit estimates. In these cases the analyst has little option. He's told what is expected of him, with no ifs, ands, or buts permitted. But it should be kept in mind that any ensuing errors or miscalculations derive from management—not from any inherent shortcoming in the statistical approach.

Competition.   Estimating what the competition will do is also to a large extent subjective. To be sure, market simulation techniques can provide solutions based on varying alternative moves by a firm and its competitors. But such assumptions must by definition be subjective because as yet no one has been able to find a surefire way of reading the minds of others. Likely actions can be assumed—but likely actions aren't always precisely the ones that are taken.

Thus, statistical sophistication is not enough to ensure accuracy. Time and time again methodology has been impeccable, but the analyst has tended to underestimate the competition—most often a competitor's promotion and advertising schemes or new or improved product introductions. No rules for avoiding such errors can be given except (1) know your industry and (2) don't underestimate what the competition can do.

This brings to mind the experience one firm had with a new product, which was far superior to others then on the market. On the basis of this superiority, the firm predicted sales amounting to 30 percent of the industry total. Unfortunately, the forecaster had underestimated the time the competition would need to come out with their own versions of the improved item. Competitors' products hit the market within a few months, and the company managed to capture only 20 percent of the market—only fractionally more than its traditional share.

Psychology.   Closely allied with the problem of evaluating the competition is the one involving consumer attitudes. All the statistical parameters (high income, high savings, low debt, etc.) may point to a jump in consumer buying, but consumers may still refuse to buy—forcing a seemingly logical forecast to fall flat on its face. An example of this occurred in early 1970. Forecasts called for sales of 10 million or more automobiles. All the determinants of car buying backed this up. Yet because of some psychological quirk, buyers failed to spend, and sales actually turned out near the 9.5-million level.

Here, however, it is difficult to blame the analyst, for the consumer can on occasion be extremely fickle. In a sense this is a factor outside the statistician's grasp, though even here progress is being made with the quantification of consumer attitude indices. To a modest extent this has helped in predicting some of the shifts in consumer buying patterns.

In addition to such indices, about the only thing that can be used to counter this persistent unknown is flexibility. If management starts with the conviction that forecasts are only a first approximation and that they will not stay put, then it can create in its organization the ability to adapt quickly to changes without traumatic readjustments and recriminations. Unsatisfactory as this may seem, flexibility of mind and of manufacturing facilities is often the best defense against the biggest of all uncertainties—the American consumer.

The Chance Factor

Statistical probability theory dictates that there is inherent variability in all numbers. The concept or assumption here is that such variation, since it is an implicit part of the data, can never be eliminated and that the analyst must learn to live with it. It is analogous to the classical coin-tossing problem, where one can never be sure of getting one head and one tail in two tosses of a coin. The tosser has to recognize that it is most likely that he will get one of each (a 50 percent probability), but he must also be prepared to accept the possibility of other chance variations of two heads (25 percent of the time) or two tails (also 25 percent of the time).

Probability theory, however, goes much beyond the task of evaluating such simple sampling errors. Most forecasting, for example, is heavily dependent on probability for meaningful interpretation. Take the problem of evaluating a forecast. The most likely result (analogous to one head and one tail) is the actual projection. The probability of any other result (analogous to two heads or two tails) is what the statistical technique aims to find out. It can best be described through a simple hypothetical example involving a trend. Assume that sales in years 1 to 5 are as shown in the following table:

Year	Sales, millions
1	$100
2	150
3	350
4	370
5	530

A mathematical trend line is drawn through these figures by the traditional method of least squares. The solid line in Figure 15-2a reveals

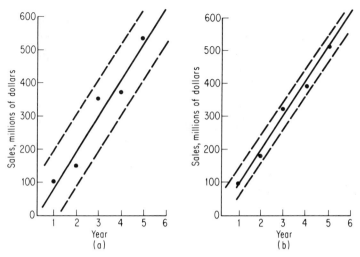

FIGURE 15-2   *Determining the statistical error.*

the results: a projection of just over $600 million for the following (sixth) year.

Next consider a series of sales which shows the following history:

Year	Sales, millions
1	$100
2	180
3	320
4	390
5	510

A trend line is fitted to these data, and the results are shown by the solid line in Figure 15-2*b*. Note that this prediction, too, is just a bit over $600 million for the following (sixth) year.

It is obvious from a look at the charts that there is lot more variation in the first example. It doesn't take any mathematical wizard to deduce that one would therefore tend to have more confidence in the second projection.

The question boils down to this: How can the analyst quantify this confidence factor or measure the likelihood that an actual sales result will depart from the projected level? Fortunately, statisticians have developed just this kind of measure, called the *standard error of estimate*. It measures the variation about the trend line, and at the same time it gives the forecaster the probability of getting such variation.

It can be shown, for example, that 1 standard error on either side of the trend-line projection should account for about two-thirds of all chance variation; 2 standard errors on either side of the trend line should account for about 95 percent of the variation; and 3 standard errors on either side of the trend line should account for virtually all possible chance errors.

This measure makes it obvious that a sales figure that is, say, 5 standard errors out could not possibly be due to chance. Such a sales projection is therefore rendered unacceptable or, to use the statistical jargon, is significantly different from the historical trend.

We come back now to our two hypothetical sales trends. By use of the proper formulas (which can be obtained from any statistical text), the following standard errors were computed:

- Example 1 (Figure 15-2a)—35
- Example 2 (Figure 15-2b)—13

Thus, in the first example one should have close to a 100 percent confidence that the prediction will fall within the range of $600 million ±3 standard errors of $600 million ($600 ± 3 × 35)—or between $495 million and $705 million. On the other hand, in the second example one can have close to 100 percent confidence that the prediction will fall within the range of $600 million ±3 standard errors of $600 million ($600 ± 3 × 13)—or between $561 million and $639 million. These ranges are shown as the dashed lines in Figure 15-2a and b. Such confidence ranges need not be limited, of course, to trend-type projections. They are widely used in correlation work and any other type of objective technique where mathematical equations are utilized. Through the use of complex statistical formulas it can be shown also that the confidence band tends to fan out as the analyst goes beyond the range of actual observation and farther out into the future.

The estimate of chance error also provides the basic rationale for statistical quality control (see Chapter 8). In this latter context, chance errors are estimated on the basis of probability formulas. And when the observed or actual errors exceed these chance limits, the production process (or whatever else is being monitored) is then said to be out of control.

Summing up, estimation of chance errors can be used for many different purposes, such as the following.

1. A measure of expected variation. The analyst is told by the standard error just how much variation he can expect on the basis of chance. Thus he knows that he must be willing, on the average, to accept a considerably larger error in the first example on page 337 than in the second one, primarily because of the wider variation inherent in the first example's data.

2. A yardstick of accuracy.   By expressing variation in terms of the standard error, the analyst can develop relative performance yardsticks. Thus if the projection were off by $35 million in the first example and by $13 million in the second one, both would be the same number of standard errors off (precisely 1)—and hence both sales forecasts would have equal relative accuracy. In this manner, projections of products or product lines with differing degrees of variability may be compared if the analyst is willing to accept a greater degree of absolute error whenever a product's past history has shown a greater amount of sales variation.

3. Warning signals.   Any time actual sales fall beyond the limits set by calculating a range based on 2 to 3 errors, the analyst is automatically warned that something is amiss. The actual sales may fall outside such control limits for just one year because of a strike or some other nonrecurring event. On the other hand, if sales fall outside such limits for several years running, then it is very likely that the past historical trend is undergoing a major shift.

4. A guide to attainable accuracy.   If nothing else, estimates of these chance errors help to point up the limits of precision. They also serve to refute the unreasonable top manager who wants every product to have the same degree of accuracy.

In forecasting sales, for example, it is clear that producers of relatively stable consumer nondurables—such as food—tend to forecast best. Thus, yearly deviations (actual versus projected) rarely exceed 4 or 5 percent. On the other hand, makers of consumer durable products, which are more sensitive to small changes in the business cycle, tend to be somewhat less accurate. In many cases large standard errors of estimates suggest deviations of 7 and 8 percent—sometimes even as much as 10 percent in a few volatile big-ticket appliance lines.

Projections of industrial components also tend to be on the low end of the accuracy scale for still another reason. That's because to forecast components, it is usually necessary to forecast demand for the products into which they are ultimately put. Sometimes this necessitates forecasts two or three rungs further along on the distribution ladder.

A maker of electronic components, for example, may have his product go to makers of fractional horsepower motors. The demand for such motors, in turn, is then dependent on the demand for small appliances. Thus in making his forecast of component sales, the electronics manufacturer has to have some idea of the sales outlook in electric motors and the small-appliance lines which use such motors. An error in any of these forecasts can easily throw off his estimate of component sales.

Attainable accuracy need not be limited to sales forecasting problems. If data are available, odds can be figured on the possible occurrence of almost any business parameter. One firm, for example, estimates the probability of many different profit levels on any new project by factoring in a whole range of estimates on costs, prices, and volume. The end result is a schedule of percent return on capital—with any given return level equated to a specific probability (see Figure 15-3). The advantage of this approach, of course, is that it gives corporate executives a summary report on the risks involved—must information for any go or no-go decision.

Evaluation

With all the known sources of error identified and, one hopes, quantified, it becomes a lot easier to evaluate the true worth of any statistical projection. Actually, several yardsticks are commonly used—ranging from the simple and naïve to the highly sophisticated.

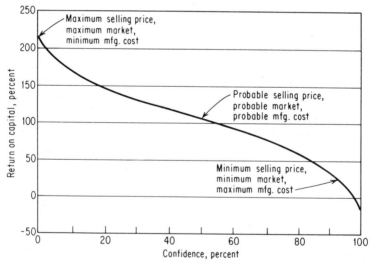

FIGURE 15-3   *Estimating profitability on an R&D project.*

To be sure, the establishment of objective quantitative criteria for evaluation does not solve all problems. Some subjective appraisal is almost always necessary. Nevertheless, quantification—since it does establish the range of error over a period or because it points up bias— adds enormously to the utility of the forecast. If nothing else, knowledge of the error configuration alerts decision makers to problem areas and points the way to corrective action.

Some of the more common of these yardsticks for evaluation are discussed below.

The absolute difference between actual results and projections. This involves little more than lining up projections with actual results and then coming up with minus or plus dollar-variance figures. While dollars in themselves are important as far as the profit and cash requirements of the firm are concerned, they can sometimes be misleading when used as the basic yardstick of accuracy. Thus a firm with $100 million in sales per year would almost certainly tend to have a larger dollar variance than, say, a firm whose sales averaged out at $100,000 per year—if only because the former company is working with bigger figures.

The percentage difference between actual results and projections. The advantage over the absolute approach described above, of course, is that this type of yardstick permits comparison when different magnitudes (sales volumes in the above example) are involved. If the smaller firm described above undershot its forecast by $1,000 and the larger by $1,000,000, both would be judged equally accurate— since actual sales of both firms would be the same 1 percent away from projected levels.

Of course, in the above example it seems pretty clear that the percentage figure is the more sensible yardstick. Obvious as this simple fact is, the author has seen it ignored time and time again, sometimes because the forecaster or the corporate executive using the projection has some particular ax to grind and other times simply because the people involved are just too close to the forecast to see it in its proper perspective.

Actual versus anticipated change. As noted above, forecasts are generally evaluated by comparing projections with actual results. But a superior test is a comparison of anticipated change with the actual variation from the latest known level. The latter is a much more sensitive measure of accuracy. Since deviations in terms of change are generally much larger, it makes it easier to distinguish between a good and a bad forecast.

The following example can best illustrate this sensitivity. If the GNP is projected to rise, say, from $1.1 trillion to $1.2 trillion but then actually rises to only $1.19 trillion, the results using both approaches are as follows: Employing the more typical actual-volume versus anticipated-volume method, the error comes out less than 1 percent ($1.19 trillion versus $1.20 trillion). But using the anticipated-change versus actual-change approach, the error is 10 percent (anticipated change of $0.10 trillion versus actual change of $0.09 trillion).

The second test affords a superior standard for evaluation since the purpose of any forecast is to anticipate as nearly as possible departures from the status quo.

The standard-error approach. This technique, described earlier in the chapter, does away with the basic objection that can be levied at all the above approaches: the fact that they don't take into account the different inherent variabilities of the data being monitored.

Thus if one firm were in a less volatile industry than a second, then by dividing observed deviations by their inherent variability (standard errors), the two could be compared on more or less equitable terms. Put another way, in the case of the more volatile second forecast, the analyst could automatically set wider limits over which the actual results could be expected to vary from projected levels on the basis of past experience.

The difference in macro versus micro errors can be resolved by this technique. Clearly, the lumping of many different lines into, say, an overall company profit forecast is likely to involve less error than if each line were taken separately. That's because, as pointed out earlier, small errors in individual series would tend to balance one another out. These compensating errors show up in terms of a lower standard error of estimate.

The standard-error approach also helps distinguish between variability of different areas of forecasting. Thus a sales forecast necessarily requires a judgment of marketing potential, which in turn rests on results of market strategies and tactics. These are most difficult to anticipate, and hence the standard error of such a sales forecast will likely be considerably larger than, say, the standard error of a forecast of the United States population trend.

Ignoring this variability factor can lead to serious errors in judgment.

Take the example of two weather forecasters. One is located in New York, where the weather patterns change almost daily. The other is located in the California desert, where it rains barely five days a year. Further assume that over a given year the New York expert, who went to meteorology school and has practiced his profession for years, is correct 80 percent of the time.

The forecaster out in the California desert, however, never took a lesson in his life. All he knows is that it seldom rains in the desert. And if he predicts "fair and warm" every single day of the year, he is bound to be right over 95 percent of the time.

If one looks merely at percentage of accuracy, one's first impulse is to say that the California man is a better forecaster. But, of course, nothing could be further from the truth; he has been relying on the lack of variability to build up his imposing record. Any city that chooses him over the New York forecaster would be making a serious mistake.

Naïve model comparison.   Another useful testing approach is to compare projected results with those obtainable from a more naïve model. Specifically, does your estimate come out significantly closer to actual results than some of the more naïve forecasting models which are already available or can be made available at considerably less cost and effort?

If the answer is "yes," then the time and money expended by the analyst in his relatively complex estimating procedures can be justified. If, on the other hand, the answer is "no," then it is advisable to think about dropping the current program—and either attempting something new or amended or going back to the simpler approaches which seem to yield almost as accurate results.

Clearly, the most naïve model to compare a forecast with is one which assumes no change from the past. This is, of course, a model which requires no advance preparation. Another model, only slightly less naïve, is to compare a firm's results with what would have been expected on the basis of simple straight-line extrapolation of the long-term trend. If profits went up 5 percent last year, this naïve model would predict the same rate of growth for the coming year. Still another popular model-comparison test is one which stacks the company's projection up against the so-called standard forecast. This is a forecast that is readily available from other analysts, either inside or outside the company.

Assume, for example, that the well-publicized consensus of economist opinion calls for auto sales in the coming year running at about 9 million units. By dint of some complex econometric calculations, your department has come up with a prediction of 9.2 million units. Under this evaluation approach, your technique would be paying off if actual sales ran at 9.3 million units, for your prediction would be more accurate than the standard one. On the other hand, the expense would be questionable if, say, actual sales turned out to be only 9.1 million units—for then there is the same 100,000-car variation in both your method and the standard method, which, in this case, was available without charge.

To sum up the naïve model-evaluation approach, a complex objective forecast is justified only if it consistently yields significantly better results than some of the simpler models and approaches readily available free of charge or at very little cost to the forecaster.

The turning-point yardstick. Another useful evaluation test involves measuring the sensitivity of a forecast or, more specifically, how well it can predict turning points. As noted at the beginning of this chapter, this ability to predict a change in direction is sometimes more important than actual dollars-and-cents accuracy. Indeed, some analysts regard this as the true test of a forecast's worth. One sales vice-president claims that he is more concerned with the projection of turning points in a sales trend—and when they will occur—than with any attempt at achieving 100 percent accuracy in predicting the actual sales volume.

Predictions of changes in direction and when they will take place, he says, "are the warning signals that the basic sales climate is changing—something that may call for a completely different kind of sales plan or promotional activity in the upcoming month, quarter, or year. Knowing when this is likely to happen is worth dollars and cents to us."

A top consultant adds that this is the area that "separates the men from the boys. It's easy to predict a 5 percent increase in sales next month—based on a similar increase last month. But it's a lot harder and takes a lot more guts to buck the prevailing trend and call for a trend reversal."

The rule of reason. Essentially, this involves evaluation on the basis of common sense. The basic question here is, Does the conclusion seem reasonable in view of what is already known? Sometimes, for example, straight extrapolation of past trends can lead to ludicrous results.

For example, if one had taken the percentage increase in the number of TV sets between 1947 and 1952 (10,000 percent) and extrapolated this percentage another five years out, one might have concluded that there would be several billion TV sets in the United States by 1957. This would have meant 10 or more TV sets for every man, woman, and child—clearly a ridiculous conclusion. There are many less obvious types of this kind of blind extrapolation. The fact is that general knowledge of an industry—industry savvy, if you will—is a must for intelligent statistical analysis and evaluation.

Subjective criteria. Quantitative measures, as pointed out before, are not the only means of evaluation. A projection stands or falls on the imperatives of a particular situation. If it serves its intended purpose, that is, if it helps anticipate change in direction, magnitude, and trend, then it can be deemed a useful forecast.

Moreover, the degree of needed accuracy can also vary depending on the company situation. Thus a firm operating on a very thin profit margin per dollar of revenue will, other things being equal, require greater accuracy in sales forecasting. A slight variation in such a case could make a difference between profit and loss.

For still other firms it may be more important to know the direction of change rather than the precise amount of change. The mix of fixed and variable costs can also be a factor in the amount of accuracy required. Less accuracy may be needed for a firm with high variable costs but low fixed costs. This situation, of course, would leave the company less vulnerable to the consequences of a marked error in sales projections.

Moreover, the degree of acceptable error would seem to depend to a large extent on the weight of any decision to which the forecast contributes. If a projection figures heavily in a decision to build or not build an expensive new plant, then it had best be accurate. Finally, the accuracy demanded should vary with the ability of the firm to respond to changes or to take corrective action in the face of poor results. A firm capable of a quick and effective response obviously requires less accuracy.

In any event, it would seem useful for every firm to set up its own accuracy yardsticks. A 3 or 4 percent off-target performance, for example, might be deemed tolerable—and anything beyond that might be used as a take-off point in search of better forecasting techniques. Indeed, many firms are doing just this—with surprisingly good results. In addition to pinpointing weak spots, it establishes a rational approach for (1) setting goals, (2) checking on results, and (3) setting in motion the machinery for necessary and useful improvement.

Systematic Checkups

Gauging a forecast's accuracy is only one part of the analyst's follow-up on the original projection. Equally if not more important are the steps taken to see that errors, if they occur, are caught quickly. Also essential is the setting up of procedures for assuring that the next forecast will be at least as good as—and preferably better than—the last one.

Such audits actually have a double-barreled payoff. On the one hand, knowledge of the error factor and/or systematic biases in the methodology adds to the utility of the forecast for business planners. On the other hand, isolation and analysis of the error factor expose weaknesses of methodology for correction or lead to the formulation of new models.

Perhaps the most obvious first step in such forecast audits is the so-called interim review. Virtually no analyst can afford to be so sure of himself that once having projected a year ahead, he rests on his laurels for the next 12 months. This was more than amply demonstrated by the researchers queried in a special 300-firm sales forecasting survey conducted by the author in 1965. Some 95 percent reported at least one review during the year. Perhaps even more impressive is the fact that nearly three out of every five analysts surveyed indicated at least a cursory check once a month.

One thing that emerges from all this talk about audit and control of forecasts is the fact that revisions are sometimes necessary, and when effected they should be regarded as another service of the forecasting group, rather than as a black mark for having erred in the first place; for in the long run, the more up to date a forecast is, the more useful it will be to the company. There certainly should be no stigma attached if the change is due to unforeseen developments in the industry or the firm. Such things sometimes happen to even the researchers of some of the nation's largest firms. A perfect illustration of this was provided a few years ago by the color-TV industry. Specifically, in early 1965 the whole industry underestimated sales and was forced to raise its sights significantly only a few months after the original forecast was issued.

At the beginning of the year the forecasters were calling for a drop-off after the usual Christmas spurt. This had always occurred in the TV industry, and the analysts saw no reason why the usual seasonal pattern wouldn't be operative again in 1965. On the basis of this reasoning, they predicted that the 145,000 units sold in December 1964 would give way to around 130,000 units in the early months of 1965.

But consumers crossed them up. Some 141,000 units were ordered in January, a hefty 76 percent ahead of levels the previous year. The same thing happened in February. The reasons for this were a sharp drop in TV prices and an increase in color programming by the major networks. Both exceeded the forecasters' expectations and helped throw the projection out of whack.

As a result, the appliance forecasters had to make some drastic revisions by the end of the first quarter of 1965. They boosted the yearly sales estimate from 2 million units to 2.2 million units. Thus, the forecasters had to make a 10 percent upward revision in a supposedly scientifically derived forecast only three months after it was first released. True, this was an unusual case, due to startling changes in key assumptions on prices and program availability. But it points up the need for a constant check on actual sales. No matter how scientific or objective a forecaster is, he has to base his projections on a certain set of assumptions. There is never any guarantee that these assumptions will hold long enough to make the forecaster look good.

From an organizational point of view, these audits can be effected in many different ways. One firm tells of getting daily reports from branch sales personnel. Armed with this information analysts keep a running check on accuracy and issue a revised forecast whenever they feel the projection is significantly and consistently departing from the actual sales figures. They also maintain constant contact with product line managers and other marketing personnel—a procedure which sometimes allows them to sense a sales change even before all the actual results are in.

Other (generally smaller) firms prefer a monthly or quarterly checkup. One analyst tells of a quarterly review where he is asked to present a report on forecasted versus actual results—along with a detailed analysis of variations which have occurred. But whatever the timing and technique, the value of audits is readily apparent: If nothing else, they provide a progress report, signaling the need for a course correction in mid-flight.

Postmortem checkups after the forecast period are equally useful. Some firms, for example, require a written appraisal at the end of the year or the forecast period. The appraisal is regarded as an integral part of the new forecast. This kind of rigorous postmortem requirement tends to stimulate thinking—to point up where the previous prediction went off the track, and generally to contribute to the continuing improvement of the forecasting procedure.

One outfit, in going over past projections, found that it was consistently underestimating summer sales. Closer examination revealed that the firm was using seasonal adjustment factors that were five years old and that these seasonals had recently undergone a change. It is doubtful whether this type of error could have been discovered without detailed checkups on past results.

Improving Statistical Intelligence

A systematic record of forecasts and a continuing review of results are only the first steps toward better statistical procedures and increased accuracy. The point to remember is that there's always room for improvement. Then, too, there's the promise of bigger payoffs. In today's competitive markets even fractional increases in accuracy can be translated into dollars-and-cents profits. Add the traditional professional yearning for perfection, and it's easy to see why statistical analysts are always looking for new ways of bettering past performance.

Some of the worthwhile questions to explore in this quest for more accuracy might include the following.

1. Can the basic data be upgraded?   In many cases errors in input data are responsible for errors in projections. A recent study of revisions in the provisional GNP estimates (upon which all business forecasters depend for estimates of general economic activity) showed that about 40 percent of the mean error in GNP forecasts was attributable to errors in the current data. This is a substantial fraction, and it represents a part of the forecasting error for which analysts per se cannot be blamed. But Uncle Sam's basic data are not the only problem. Much of the initiative to improve upon the raw statistics must in the end come from the analyst. Statisticians, as a general rule, should always be searching for opportunities to upgrade the basic data, either by finding new or more accurate sources or by investigating the possibility of getting needed information a bit sooner.

On the latter score, one company tells of having its trade association submit industry sales estimates to the firm one week earlier than they were released for general consumption. This was only a short period of time, and yet it was enough to allow the company, in preparing its short-term forecast, to have the benefit of one month's additional data—enough to result in a marked improvement in sales forecasting accuracy.

Data procurement improvements needn't stop at that point. Many firms, before they use any specific statistical series, make thorough checks on its coverage and accuracy. Too often, there's the temptation to accept published figures at face value—based on the fallacy that "if it's in print, it's probably accurate."

2. Are your approaches tailored to your functions and markets? One forecast will not serve all areas of management. For example, a forecast used in financial planning is characteristically long-range, dealing principally in aggregate figures. In promotion planning, the forecast is not usually long-range; the performance to date and projections to the end of the season by model and style are the common guides.

In production inventory control, the forecast requires detailed information, not only on model but also on colors and sizes within the model. Although the need for detail is greater, the forecasting period is considerably shorter. The immediate future is important: the next few weeks or the next few months.

Similarly, just as one forecast is not sufficient for all manufacturing operations, no single technique applies to all marketing situations. Successful forecasting with a particular method in one market tells little or nothing about its potential success in another market.

3. Can your techniques be improved upon? The advent of the computer has opened up vast new potentials to all firms. No longer need they be wedded to the old-fashioned, simple correlation or trendline technique. Some of the more subtle approaches such as step regression and model building are within the reach of almost every firm.

Similarly, the computer allows for the testing of many additional variables. Experimentation, which was costly by hand, is almost routine when machines are made available.

A large consumer hard-goods firm tells how it tested almost a hundred variables before it came up with the best combination. The cost? A few thousand dollars. The result? A decline in the average error from 6 percent way down to 2 percent in the first two years after the implementation of the new system.

But one needn't have a computer to improve on forecasting

techniques. One has only to read the technical journals to discover the spate of new approaches or variations on old ones that come out in a seemingly unending procession. Surely one or two of these can be applied or at least test-fitted to an individual firm's forecasting problems.

Many companies periodically review their statistical procedures in the light of the improved state of the art. One marketer sums it all up: "If we don't improve our forecasting techniques, we can be sure our competitors will. We just can't afford to coast along in this business."

4. Are the assumptions realistic?   It has been brought out above that changing assumptions can lead to serious errors. Forecasting is built on a foundation of assumptions—about the economy, about the industry, about competing firms, and about what the individual firm can and cannot do. Thus it is usually a good idea to sit down once in awhile and evaluate the entire spectrum of assumptions in the hope that perhaps a more realistic approach to some may yield a more accurate sales projection.

5. Is your forecast being integrated?   The most accurate long- or short-range sales forecast is of little value unless it becomes an integrated part of an information system on which decisions are based. Information systems used in such areas as financial planning, promotion, activity planning, and production-inventory control rely heavily on a sales forecast as their foundation.

6. Are your approaches practical?   The most accurate projection is one which considers each single relevant factor. But the practical problems of money and time limit one to a narrower choice. The trick, of course, is to choose those specific variables which will yield the most dollars-and-cents return. Thus the analyst has to wear two hats—that of a hardheaded businessman and that of the ivory-towered theoretician.

A related problem stems from the fact that there is a large gulf between a theoretically sound technique and a successful, practical application. Beyond the textbook, one must add intuitive judgment, the analysis of error, and the stability of product and market.

7. Are you using all available statistical aids?   Statisticians over recent years have developed scores of shortcuts and laborsaving techniques—all designed to take the drudgery out of calculations and reduce the incidence of clerical error. Here are just a few that can make your work considerably easier and cut down on the chances of serious error:

(a) TABLES: Almost every standard statistical text contains tables which list in easy-to-read column form the results of computations that occur over and over again. Thus, there are tables of squares, square roots, cubes, cube roots, the probability of an event's occurring, etc.

Some tables have many uses. For example, the common compound interest rate tables which appear in almost any business arithmetic text can also be used to estimate growth rates. All that's necessary is the substitution of the term "growth rate" for the term "interest rate" in the table.

(b) FORMULAS: In some instances formulas are available that are expressly designed to reduce calculations. If, for example, you had to sum up the squares of a series of consecutive numbers, you could get the result by using a simple algebraic formula—thereby eliminating a lot of useless multiplication (squaring) and addition.

(c) LOGARITHMS: As noted in Chapter 3, the concept of logs is extremely useful in statistical analysis. In addition to having graphic advantages, it can cut down the complexity of calculations to a considerable degree. The principles of logarithmic theory are surprisingly simple and can usually be fully appreciated in a matter of an hour or two. It's usually well worth the effort because the use of logs eliminates many time-consuming computations involving multiplication and division.

(d) OTHER AIDS: Many other shortcuts are also available to the statistical worker, depending upon the actual problem under study. For example, there are innumerable techniques for manipulating formulas— putting them into a form adapted to today's high-speed mechanical calculators.

Another important aid at the disposal of the statistical analyst is the check. Many computations in statistics are set up so that by merely adding another column to your work sheet, you get a complete check on numerical accuracy. Such checks also help pinpoint where an error may have occurred.

8. *Do you capitalize on errors?* Even a poor formal forecast is better than none at all. A forecast error, in fact, gives useful data. When properly quantified and incorporated into your overall control system, it can help determine such things as required safety inventory or allowable off-season production.

Capitalizing on errors also involves the periodic checkups and postmortems discussed on page 346. Zeroing in on where he went wrong can perhaps enable the forecaster to avoid some of the pitfalls the next time around—and, one would hope, to come up with more reliable results.

9. *Are you getting top mileage from your statistical efforts?*
Many firms insist on periodic appraisals of their statistical programs
to determine their usefulness to the company's operating divisions. Too
often forecasts tend to be ignored or not presented in a form in which
they can be readily adopted for line personnel usage.

One marketing vice-president notes that it took almost six months
of intensive educational effort on his part to convince sales and other
divisions that the corporate sales forecast was something more than one
of top management's newfangled ideas—that it was a useful blueprint
for line planning and actual day-to-day operation. After the indoctri-
nation the company's quarterly sales outlook became one of the most
eagerly awaited reports in the entire firm; some units called up if it
was as much as one day late.

Blame for lack of proper forecast usage can't always be laid at the
door of the receiving party. One company executive tells of the problem
of having to revamp the whole forecasting setup to make the predictions
compatible with existing market conditions. Originally, the forecasts were
being prepared on a state-by-state basis since income (the major determinant of this firm's sales) had traditionally been reported in this way.
Unfortunately, the company products were being marketed by metropolitan area—and the income trends of states didn't always correspond
to those of marketing areas. In other words, the state-oriented forecasts
weren't proving too useful.

A thorough reappraisal resulted in a complete revamping of the sales
forecasting report. With the help of additional money and personnel,
the firm reworked income statistics on a metropolitan-area basis. Expensive? Yes, but the new approach has paid for itself many times over
in the form of more useful sales forecasts.

There is an additional bonus to the sales forecaster who can get the
operating divisions to use his sales projection. Other things being equal,
use tends to enhance accuracy. For the sales people, given a blueprint
of the most likely sales trend, have received valuable sales intelligence—
the kind they can use to realistically pursue and hence achieve company
sales objectives.

10. *Is the forecast an attainable one?*   Sometimes a forecast can
be statistically perfect, but just doesn't mesh with existing conditions
within the firm. One manufacturer explains that a forecast called for
selling about 100,000 units of a machine component. This forecast had
to be toned down simply because the firm did not have production

facilities for such volume and could not build new ones in time to meet the forecast level.

In a similar vein, the forecaster must consider such other internal factors as the financial position of the firm. The firm might be able to sell more, but it might be overextending itself by taking on too heavy a financial load in terms of working capital and accounts receivable. In still another area, a firm might be willing to boost sales substantially, but might just not have the time to train a competent sales staff.

Under the subject of attainability some people also list supply availability. In 1965, for example, several brass mills fell short of their sales targets—not because they could not sell additional products, but simply because they could not get the necessary copper, which was then in short supply.

11. Where should improvements be made? Growing product and market complexities may warrant an upgrading of grass-roots divisional forecasting. But given limited resources, it is unlikely that improvement can be achieved for all products and markets simultaneously. Where, then, to concentrate? The following are some key questions that might be explored:

(a) IS THE EFFORT WORTHWHILE? The volume must be sufficient to justify both management's attention and the cost of sophisticated econometric analysis.

(b) HOW DOES PAST PERFORMANCE RATE? Aim at weak areas—say, where a division has been caught long on inventory in a downturn or short on productive capacity during an upturn.

(c) IS A QUANTITATIVE APPROACH FEASIBLE? This presupposes adequate statistical history and product responsiveness to economic causative factors.

(d) IS DIVISIONAL MANAGEMENT RECEPTIVE? Doubting top brass can scuttle the best of forecasting techniques.

Supplementary Reading List

Abramson, Adolph G., and Russell H. Mack: *Business Forecasting in Practice: Principles and Cases,* Wiley, New York, 1956.

Adams, Joe Kennedy: *Basic Statistical Concepts,* McGraw-Hill, New York, 1955.

Adler, Irving: *Probability and Statistics for Everyman: How to Understand and Use the Laws of Chance,* John Day, New York, 1963.

Aigner, Dennis J.: *Principles of Statistical Decision Making,* Macmillan, New York, 1968.

Allan, Douglas H.: *Statistical Quality Control,* Reinhold, New York, 1959.

Bailey, Richard M.: "Tailoring the Business Forecast to Company Size," *Business Horizons,* Summer 1962.

Balsey, Howard L.: *Introduction to Statistical Method, with Applications to Commercial, Economic and Industrial Data,* Littlefield, Adams, Paterson, N.J., 1964.

Bassie, V. Lewis: *Economic Forecasting,* McGraw-Hill, New York, 1958.

Bernstein, Leonard A.: *Statistics for Decisions: A Tool for Everybody,* Grosset & Dunlap, New York, 1965.

Blackwell, David: *Basic Statistics,* McGraw-Hill, New York, 1969.

Blackwell, David, and M. A. Gershick: *Theory of Games and Statistical Decisions,* Wiley, New York, 1954.

Bopp, Eberhard: "Better Short-term Planning through Computerized Time Series Forecasting," speech before the Sixth Annual Forecasting Conference, American Statistical Association, New York, Apr. 17, 1964.

Bratt, Elmer C.: *Business Forecasting,* McGraw-Hill, New York, 1958.

Bryant, Edward C.: *Statistical Analysis,* 2d ed., McGraw-Hill, New York, 1966.

Buzzell, Robert D., and Charles C. Slater: "Decision Theory and Marketing Management," *Journal of Marketing,* vol. 27, July 1962.

Chao, Lincoln L.: *Statistics: Methods and Analysis,* McGraw-Hill, New York, 1969.

Chartner, William: "Using and Evaluating Forecasts," speech before the National Association of Business Economists Seminar on Business Forecasting, December 1963.

Chase, Clinton I.: *Elementary Statistical Procedures,* McGraw-Hill, New York, 1967.

Clelland, Richard C.: *Basic Statistics with Business Applications,* Wiley, New York, 1966.

Cole, Harrison W.: "Special Problems of Seasonally Adjusting Company Data," speech before the American Statistical Association, September 1963.

Cowden, Dudley J.: *Statistical Methods in Quality Control,* Prentice-Hall, Englewood Cliffs, N.J., 1957.

Cox, Clifton B.: "Using Computers in Pricing," speech before the Tenth Annual Marketing Conference, National Industrial Conference Board, Sept. 20, 1962.

Craddock, J. M.: *Statistics in the Computer Age,* English Universities Press, London, 1968.

Crawford, C. M.: *Sales Forecasting: Methods of Selected Firms,* University of Illinois Press, Urbana, 1955.

Crowe, Walter E.: *Index Numbers: Theory and Applications,* Macdonald & Evans, London, 1965.

Croxton, Frederick E., and Dudley J. Cowden: *Applied General Statistics,* 3d ed., Prentice-Hall, Englewood Cliffs, N.J., 1967.

D'Arcy, J. A.: *Chance and Choice: Practical Probability and Statistics,* Thames & Hudson, London, 1968

Dixon, Wilfrid J., and Frank J. Massey: *Introduction to Statistical Analysis,* 3d ed., McGraw-Hill, New York, 1968.

Dubois, Edward: *Essential Methods in Business Statistics,* McGraw-Hill, New York, 1964.

Epstein, Richard A.: *The Theory of Gambling and Statistical Logic,* Academic, New York, 1967.

Ferber, Robert: *Statistical Techniques in Market Research,* McGraw-Hill, New York, 1949.

Forecasting in Industry, Studies in Business Policy, no. 77, National Industrial Conference Board, New York, 1956.

Forecasting Sales, Studies in Business Policy, no. 106, National Industrial Conference Board, New York, 1963.

Forester, John: *Statistical Selection of Business Strategies,* Irwin, Homewood, Ill., 1968.

Frank, Ronald: *Quantitative Techniques in Marketing,* Irwin, Homewood, Ill., 1962.

Fruend, John E., and Frank J. Williams: *Modern Business Statistics,* Prentice-Hall, Englewood Cliffs, N.J., 1958.

Goldfarb, Nathan, and William K. Kaiser: *Gantt Charts and Statistical Quality Control: The Dissemination of New Business Techniques,* Hofstra University, Hempstead, N.Y., 1964.

Goodman, Oscar R.: *Sales Forecasting,* Wisconsin Commerce Reports, vol. 3, no. 5, May 1954.

Grant, Eugene L.: *Statistical Quality Control,* McGraw-Hill, New York, 1946.

Graybill, Franklin A.: *An Introduction to Linear Statistical Models,* McGraw-Hill, New York, 1961, vol. I.

Hadley, George: *Introduction to Business Statistics,* Holden-Day, San Francisco, 1968.

Hastings, Delbert C.: *Forecasting in Small Business Planning,* Management Research Summary, Small Business Administration, April 1961.

Hein, Leonard W.: *The Quantitative Approach to Managerial Decisions,* Prentice-Hall, Englewood Cliffs, N.J., 1967.

Jaszi, George: "Forecasting with Judgmental and Econometric Models," speech before the Seventh Annual Forecasting Conference, American Statistical Association, New York, December 1963.

Kaplan, Lawrence Jay: *Elementary Statistics for Economics and Business,* Pitman, New York, 1966.

Karchere, A. J.: "Computer Applications in Business Forecasting," speech before the American Statistical Association, New York, December 1963.

Kazmier, Leonard J.: *Statistical Analysis for Business and Economics,* McGraw-Hill, New York, 1967.

King, William R.: *Probability for Management Decisions,* Wiley, New York, 1968.

Levin, Richard I., and Rudolph P. Lamone: *Linear Programming for Management Decisions,* Irwin, Homewood, Ill., 1969.

Lewis, Edward Erwin: *Methods of Statistical Analysis in Economics and Business,* Houghton Mifflin, Boston, 1963.

Lorie, James H.: "Two Important Problems in Sales Forecasting," *Journal of Business,* July 1957.

McLaughlin, R. L.: *Times Series Forecasting,* Marketing Research Techniques, no. 6, American Marketing Association, 1962.

Mandel, Benjamin J.: *Statistics for Management: A Simplified Introduction to Statistics,* Dongary, Baltimore, 1964.

Matamoros, Albert G.: *The Check and Balance Forecasting Process at Armstrong,* National Association of Business Economists, September 1963.

Materials and Methods of Sales Forecasting, Special Report 27, American Management Association, New York, 1957.

Mills, Frederick C.: *Statistical Methods,* 3d ed., Holt, New York, 1955.

Moore, Peter Gerald: *Statistics and the Manager: The Use of Statistics and Probability in Managerial Decisions,* MacDonald, London, 1966.

Morgan, Bruce W.: *An Introduction to Bayesian Statistical Decision Processes,* Prentice-Hall, Englewood Cliffs, N.J., 1968.

Morgenstern, Oskar: *On the Accuracy of Economic Observations,* 2d ed., Princeton, Princeton, N.J., 1963.

Morris, William T.: *The Analysis of Management Decisions,* Irwin, Homewood, Ill., 1964.

Mueller, Robert K.: *Effective Management through Probability Controls: How to Calculate Managerial Risks,* Funk & Wagnalls, New York, 1950.

Murphy, Frank: "New Perspectives of Appliance Forecasting," speech before the Sixth Annual Forecasting Conference, American Statistical Association, New York, Apr. 17, 1964.

O'Hara, John Brangs: *Effective Use of Statistics in Accounting and Business: Illustrative Cases,* Holt, New York, 1964.

Oliver, Francis R.: *What Do Statistics Show?* Hodder, London, 1964.

Paden, Donald W., and E. F. Lindquist: *Statistics for Economics and Business,* 2d ed., McGraw-Hill, New York, 1956.

Parkany, John: "A New Approach to Sales Forecasting and Production Scheduling," *Journal of Marketing,* January 1961.

Peach, Paul: *Quality Control for Management,* Prentice-Hall, Englewood Cliffs, N.J., 1964.

Quality and Economic Significance of Anticipation Data, Princeton, Princeton, N.J., 1960.

Reichard, Robert S.: *Practical Techniques of Sales Forecasting,* McGraw-Hill, New York, 1966.

Reichmann, William: *Use and Abuse of Statistics,* Oxford, New York, 1962.

Riggs, James L.: *Economic Decision Models for Engineers and Managers,* McGraw-Hill, New York, 1968.

Sales Forecasting: Uses, Techniques, and Trends, Special Report 16, American Management Association, New York, 1956, pp. 148–149.

Sasaki, Kyohei: *Statistics for Modern Business Decision Making,* Wadsworth, Belmont, Calif., 1968.

Schlaifer, Robert: *Introduction to Statistics for Business Decisions,* McGraw-Hill, New York, 1961.

Schrock, Edward M.: *Quality Control and Statistical Methods,* 2d ed., Reinhold, New York, 1957.

Shao, Stephen P.: *Statistics for Business and Economics,* F. Merril Books, Columbus, Ohio, 1967.

Snedecor, George W.: *Everday Statistics: Facts and Fallacies,* Wm. C. Brown Company Publishers, Dubuque, Iowa, 1950.

Spear, Mary E.: *Practical Charting Techniques,* McGraw-Hill, New York, 1969.

Spurr, William A., and Charles P. Bonini: *Statistical Analysis for Business Decisions,* Irwin, Homewood, Ill., 1967.

Spurr, William A., Lester S. Kellogg, and John H. Smith: *Business and Economic Statistics,* 2d ed., Irwin, Homewood, Ill., 1961.

Stigler, George, and James K. Kindahl: *The Behavior of Industrial Prices,* National Bureau of Economic Research, New York, 1970.

Stockton, John R.: *Introduction to Business and Economic Statistics,* South-Western Publishing Company, Cincinnati, 1966.

Strong, Lydia: "Sales Forecasting Problems, and Prospects," *Management Review,* September 1956.

Teitelman, Samuel: *The Place for Assumptions and Judgments in Sales Forecasting,* Management Report 15, American Management Association, New York, 1958.

Terborgh, George: *Business Investment Policy,* Machinery and Allied Products Institute, Washington, 1958.

Thek, Robert R.: "Sales Forecasting of New Products," speech before the American Statistical Association, New York, Apr. 26, 1963.

Tuttle, Alva M.: *Elementary Business and Economic Statistics,* McGraw-Hill, New York, 1957.

Vesselo, Isaac R.: *How to Read Statistics,* Harrap, London, 1962.

Wallis, Wilson Allen, and Harry V. Roberts: *The Nature of Statistics,* Collier Books, New York, 1962.

Weld, Walter E.: *How to Chart: Facts from Figures with Graphs,* Codex Book Co., Norwood, Mass., 1959.

Williams, N.: *Linear and Non-linear Programming in Industry,* Sir Isaac Pitman & Sons, London, 1967.

Winters, Peter R.: "Forecasting Sales by Exponentially Weighted Moving Averages," *Management Science,* April 1960.

Zarnowitz, Victor.: "Forecasting Business Conditions: A Critical View," Selected Papers, no. 15, University of Chicago Press, 1964.

Zeisel, Hans: *Say It with Figures,* 4th ed., Harper & Row, New York, 1957.

Index

7 DAY USE

RETURN TO